Pauli Lectures on Physics Volume 6

Selected Topics in Field Quantization

Wolfgang Pauli

Edited by Charles P. Enz

Translated by S. Margulies and H. R. Lewis

Foreword by Victor F. Weisskopf

DOVER PUBLICATIONS, INC.
Mineola, New York

Bibliographical Note

This Dover edition, first published in 2000, is an unabridged republication of the work originally published in 1973 by The MIT Press, Cambridge, Massachusetts and London, England.

Library of Congress Cataloging-in-Publication Data

Pauli, Wolfgang, 1900–1958.
[Feldquantisierung. English]
Selected topics in field quantization / Wolfgang Pauli ; edited by Charles P. Enz ; translated by S. Margulies and H.R. Lewis ; foreword by Victor F. Weisskopf.
p. cm. — (Pauli lectures on physics ; v. 6)
Originally published: Cambridge, Mass. : MIT Press, 1973.
Includes bibliographical references and index.
ISBN 0-486-41459-0 (pbk.)
1. Quantum field theory. I. Enz, Charles P. (Charles Paul), 1925– II. Title.

QC3 .P35 2000 vol. 6
[QC174.45]
530 s—dc21
[530.14'3]

00-031581

Manufactured in the United States of America
Dover Publications, Inc., 31 East 2nd Street, Mineola, N.Y. 11501

Pauli Lectures on Physics in Dover Editions

Volume 1. Electrodynamics (41457-4)
Volume 2. Optics and the Theory of Electrons (41458-2)
Volume 3. Thermodynamics and the Kinetic Theory of Gases (41461-2)
Volume 4. Statistical Mechanics (41460-4)
Volume 5. Wave Mechanics (41462-0)
Volume 6. Selected Topics in Field Quantization (41459-0)

Contents

Foreword

It is often said that scientific texts quickly become obsolete. Why are the Pauli lectures brought to the public today, when some of them were given as long as twenty years ago? The reason is simple: Pauli's way of presenting physics is never out of date. His famous article on the foundations of quantum mechanics appeared in 1933 in the German encyclopedia *Handbuch der Physik*. Twenty-five years later it reappeared practically unchanged in a new edition, whereas most other contributions to this encyclopedia had to be completely rewritten. The reason for this remarkable fact lies in Pauli's style, which is commensurate to the greatness of its subject in its clarity and impact. Style in scientific writing is a quality that today is on the point of vanishing. The pressure of fast publication is so great that people rush into print with hurriedly written papers and books that show little concern for careful formulation of ideas. Mathematical and instrumental techniques have become complicated and difficult; today most of the effort of writing and learning is devoted to the acquisition of these techniques instead of insight into important concepts. Essential ideas of physics are often lost in the dense forest of mathematical reasoning. This situation need not be so. Pauli's lectures show how physical ideas can be presented clearly and in good mathematical form, without being hidden in formalistic expertise.

Pauli was not an accomplished lecturer in the technical sense of the word. It was often difficult to follow his courses. But when the sequence of his thoughts and the structure of his logic become apparent, the attentive follower is left with a new and deeper knowledge of essential concepts and with a clearer insight into the splendid architecture of reason, which is theoretical physics. The value of the lecture notes is not diminished by the fact that they were written not by him but by some of his collaborators. They bear the mark of the master in their conceptual structure and their mathematical rigidity. Only here and there does one miss words and comments of the master. Neither does one notice the passing of time in his lectures, with the sole exception of the lectures on field quantization, in which some concepts are formulated in a way that may appear old-fashioned to some today. But even these lectures should be of use to modern students because of their compactness and their direct approach to the central problems.

May this volume serve as an example of how the concepts of theoretical physics were conceived and taught by one of the great men who created them.

Victor F. Weisskopf

Cambridge, Massachusetts

Preface

"Feldquantisierung" (the title of the German original of these lectures) has become a widely used document of the field theoretic literature from the moment it was published. The course, given by Pauli at ETH, Zürich, in 1950–51, took the form of a research seminar in which Pauli commented on the problems of current interest at the time. In this respect these lectures differ from the others of this series, which were of more conventional form and addressed themselves to the diploma students of ETH.

The character of comments is also reflected in the style of the notes, which were prepared by the late M. R. Schafroth, at that time Pauli's assistant. Coauthor of these notes was U. Hochstrasser, now delegate for scientific matters to the Swiss Government. But this style is not only characteristic of the form Pauli gave to these lectures; it also reflects Schafroth's personal manner of talking physics. In fact, the same no-nonsense style is apparent in the lectures on statistical mechanics, in this series, which are also translated from notes by Schafroth.

Of course, field theory has come a long way since the days of "Feldquantisierung." The axiomatic approach of modern field theory has introduced a quite different mathematical language and a degree of rigor without which the known exact results would not have been possible. In order to make some connection with these recent developments I have asked Klaus Hepp to comment in the appendix on the present state of the problem of renormalization without perturbation theory, and I here acknowledge his expert help.

The physical problems, though, have not fundamentally changed, except for having become more complex through the multiple discoveries in elementary particle physics. For this reason Pauli's comments on these problems in "Feldquantisierung" are still of intrinsic interest today. Apart from this the present lectures have acquired an interest as an historical document on the heroic times of quantum electrodynamics as symbolized by the names of Tomonaga, Schwinger, Feynman, Dyson, and others. For this reason I have supplemented all the casual references to original works made in "Feldquantisierung" with their exact location in the published literature. In this task I have enjoyed the encouraging help of Freeman Dyson. I also acknowledge a helpful correspondence with R. Glauber.

The editing of this English translation has not been an easy task, and the work of the translators deserves special mention. At some places I have dared to deviate from the original by adding more precision. Where this was not possible by minor changes, a comment has been added in the appendix. In this task as well as in the elimination of errors the comments by Barry Simon, at the time a graduate student of A. S. Wightman and now a noted field theorist in his own right, were of great help.

If in editing these lectures a book of interest to today's students and researchers in field theory has been produced, it will contribute to honor the memory of Wolfgang Pauli.

Charles P. Enz

Geneva, 27 October 1971

Selected Topics in Field Quantization

PAULI LECTURES ON PHYSICS VOLUME 6

Chapter 1. Quantization of the Electron-Positron Field

1. THE HEISENBERG AND INTERACTION REPRESENTATIONS [A-1][1]

We can choose the time dependence of the operators and eigenfunctions quite arbitrarily as long as the expectation values preserve the correct time dependence:

$$\langle A\rangle = \sum_{nm} (\Psi_n^* A_{nm} \Psi_m) \, . \tag{1.1}$$

1. *Heisenberg representation.* The state vector Ψ is time independent. A satisfies the field equations with interaction. For example, for $A \to \Phi_\mu$, the electromagnetic potential,

$$\Box \Phi_\mu = -j^\mu \, . \tag{1.2}$$

2. *Interaction representation.* Here Ψ is time dependent, but such that the resulting time dependence of A has the consequence that A satisfies the field equations without interaction. For example,

$$\Box \Phi_\mu = 0 \, . \tag{1.3}$$

The two representations are the same when there is no interaction. For example, if the interaction is

$$H = -j^\mu \Phi_\mu \, , \qquad j^\mu \propto e \, , \qquad e \ll 1 \, ,$$

and one expands

$$\Psi = \Psi_0 + e\,\Psi_1 + \dots ,$$

[1] Comments [A-1]–[A-7] appear in the Appendix on pp. 177–181.

then Ψ_0 is time independent and Ψ_1, Ψ_2, Ψ_3, ... are time dependent.

Note: 1. All of this is valid for ordinary quantum mechanics as well as for quantized field theories.
2. If the total energy is diagonal, then in the Heisenberg representation

$$A_{nm}(t) = A_{nm}(0)\cdot\exp[i(E_n - E_m)t] ,$$

and Ψ is time independent.

3. *Schrödinger representation.* Here A is chosen to be time independent and, accordingly,

$$\Psi_n(t) = \Psi_n(0)\cdot\exp[iE_n t] \tag{1.4}$$

(if the total energy is diagonal).

2. QUANTIZATION OF THE HARMONIC OSCILLATOR

The Hamiltonian is

$$H = \frac{1}{2}\left(\frac{p'^2}{m} + m\omega^2 q'^2\right).$$

Let

$$\frac{p'}{\sqrt{m}} = p\,, \qquad q'\cdot\sqrt{m} = q$$

(canonical transformation; p, q Hermitian). Then,

$$H = \tfrac{1}{2}(p^2 + \omega^2 q^2)\,, \tag{2.1}$$

$$i[p, q] = 1\,. \tag{2.2}$$

We introduce

$$\left.\begin{aligned} a &= \frac{1}{\sqrt{2\omega}}(p - i\omega q) \\ a^* &= \frac{1}{\sqrt{2\omega}}(p + i\omega q) \end{aligned}\right\}, \tag{2.3}$$

so that

$$[a, a^*] = 1\,. \tag{2.4}$$

Then,

$$H=\frac{\omega}{2}(aa^*+a^*a)=\omega\left(a^*a+\frac{1}{2}[a, a^*]\right). \qquad [2.5]$$

The term $\frac{1}{2}[a, a^*]=\frac{1}{2}$ is the zero-point energy. The quantity a^*a has integral eigenvalues:

$$a^*a=N \qquad (N=0, 1, 2, \ldots), \qquad [2.6]$$

which follows from the Hermiticity requirement for p and q. The matrix representation in which N is diagonal is

$$\left.\begin{array}{l} a=\begin{pmatrix} 0 & \sqrt{1} & 0 & \cdots & 0 & \cdots \\ 0 & 0 & \sqrt{2} & \cdots & 0 & \cdots \\ \cdots & \cdots & \cdots & \cdots & \cdots & \cdots \\ 0 & 0 & 0 & \cdots & \sqrt{N} & \cdots \\ \cdots & \cdots & \cdots & \cdots & \cdots & \cdots \end{pmatrix} \\ a^*=\begin{pmatrix} 0 & 0 & 0 & \cdots & 0 & \cdots \\ \sqrt{1} & 0 & 0 & \cdots & 0 & \cdots \\ 0 & \sqrt{2} & 0 & \cdots & 0 & \cdots \\ \cdots & \cdots & \cdots & \cdots & \cdots & \cdots \\ 0 & 0 & 0 & \cdots & \sqrt{N} & \cdots \\ \cdots & \cdots & \cdots & \cdots & \cdots & \cdots \end{pmatrix} \\ N=\begin{pmatrix} 0 & 0 & 0 & \cdots & 0 & \cdots \\ 0 & 1 & 0 & \cdots & 0 & \cdots \\ 0 & 0 & 2 & \cdots & 0 & \cdots \\ \cdots & \cdots & \cdots & \cdots & \cdots & \cdots \\ 0 & 0 & 0 & \cdots & N & \cdots \\ \cdots & \cdots & \cdots & \cdots & \cdots & \cdots \end{pmatrix} \end{array}\right\}. \qquad [2.7]$$

Let Ψ be a function of the variables N: $\Psi=\Psi(N)$. Then, the significance of a is seen to be as follows:

$$\left.\begin{array}{l} a^* \text{ is a creation (emission) operator, since} \\ \qquad a^*\Psi(N)=\sqrt{N+1}\,\Psi(N+1)\,; \\ a \text{ is an annihilation (absorption) operator, since} \\ \qquad a\Psi(N)=\sqrt{N}\,\Psi(N-1) \end{array}\right\}. \qquad [2.8]$$

The state of lowest energy is $N=0$, the "vacuum." Then,

$$\left.\begin{aligned} a^*\Psi(0) &= \Psi(1)\,, & a\Psi(0) &= 0\,, \\ \langle a^*a\rangle_0 &= 0\,, & \langle aa^*\rangle_0 &= 1 \end{aligned}\right\}. \qquad [2.9]$$

The occupation number N is arbitrary here; this quantization thus corresponds to Bose-Einstein statistics. Corresponding relations exist for Fermi statistics (satisfying the exclusion principle).

Formally, if one introduces

$$\{a, a^*\} \equiv aa^* + a^*a = 1\,, \qquad a^2 = 0\,, \qquad a^{*2} = 0\,, \qquad [2.10]$$

then one obtains as solution

$$a = \begin{pmatrix} 0 & 1 \\ 0 & 0 \end{pmatrix}, \qquad a^* = \begin{pmatrix} 0 & 0 \\ 1 & 0 \end{pmatrix}. \qquad [2.11]$$

Furthermore, if we set

$$N \equiv a^*a = \begin{pmatrix} 0 & 0 \\ 0 & 1 \end{pmatrix}, \qquad [2.12]$$

then

$$1 - N = aa^* = \begin{pmatrix} 1 & 0 \\ 0 & 0 \end{pmatrix},$$

$$N(1-N) = 0\,. \qquad [2.13]$$

This corresponds exactly to the exclusion principle. Note that here, in contrast with Bose-Einstein statistics, complete symmetry exists between a and a^*, and N and $1-N$, respectively.

3. SECOND QUANTIZATION FOR SPIN-$\frac{1}{2}$ PARTICLES

a. *Nonrelativistic formulation for spin-0*

We expand in terms of a complete set of eigenfunctions,

$$\psi(x, t) = \sum_r a_r \exp\left[i(\boldsymbol{k}_r \cdot \boldsymbol{x}_r - k_r^0 t)\right], \qquad [3.1]$$

and require that the amplitudes of the individual modes commute and that each individual mode behave as a harmonic oscillator:

$$[a_r, a_s] = [a_r^*, a_s^*] = 0 ; \qquad [a_r, a_s^*] = \delta_{rs} . \tag{3.2}$$

If we imagine the whole system to be enclosed within a box G whose sides are of length L, then

$$k_r^i = \frac{2\pi}{L} s^i , \quad i = 1, 2, 3 , \quad \text{where the } s^i \text{ are integers.} \tag{3.3}$$

Instead of enclosing the system in such a box, one can, with the same result, require periodicity:

$$\psi(x^1 + L, x^2, x^3; t) = \gamma \cdot \psi(x^1, x^2, x^3; t) ; \qquad |\gamma|^2 = 1 . \tag{3.4}$$

The completeness relation demands that

$$\int_G \psi^* \psi \, d^3x = \sum_r a_r^* a_r = \sum_r N_r . \tag{3.5}$$

b. Relativistic formulation

Here a characteristic complication appears, which results from the fact that solutions of all simple field equations contain negative as well as positive frequencies.

In particular, consider the Dirac equation:

$$\gamma^\mu \gamma^\nu + \gamma^\nu \gamma^\mu = 2\delta_{\mu\nu} , \tag{3.6}$$

$$\left(\gamma^\nu \frac{\partial}{\partial x^\nu} + m \right) \psi = 0 . \tag{3.7}$$

As is well known, this equation also leads to solutions with positive and negative frequencies.

We expand ψ in terms of a complete set of eigenfunctions,

$$\psi_\varrho = \sum_r A_r u_\varrho^{(r)}(x) , \tag{3.8}$$

where x is a four-vector ($x^0 = t$, $x^4 = it$), and normalize the

$u_\varrho^{(r)}$ according to

$$\int \sum_{\varrho} u_\varrho^{*(r)} u_\varrho^{(s)} \, d^3x = \delta_{rs} \, . \qquad [3.9]$$

This is possible since the conservation law for the current which follows from Eq. [3.7] guarantees the time constancy of the integral in Eq. [3.9]. As proof, we consider the adjoint equation (where the arrow indicates that differentiation acts on the left)

$$\bar{\psi}\left(\gamma^\nu \frac{\overleftarrow{\partial}}{\partial x^\nu} - m\right) = 0 \, , \qquad [3.10]$$

and form the sum of $\bar{\psi}$ times Eq. [3.7] and Eq. [3.10] times ψ:

$$\bar{\psi}\left(\gamma^\nu \frac{\partial}{\partial x^\nu} + m\right)\psi + \bar{\psi}\left(\gamma^\nu \frac{\overleftarrow{\partial}}{\partial x^\nu} - m\right)\psi = 0 \, .$$

We obtain

$$\frac{\partial j^\nu}{\partial x^\nu} = 0 \, , \qquad [3.11]$$

where (with an arbitrary constant C)

$$j^\nu = C \cdot \bar{\psi}\gamma^\nu\psi \, . \qquad [3.12]$$

From this it follows that

$$\frac{\partial}{\partial t}\int j^0 \, d^3x = 0 \, . \qquad \text{Q.E.D.}$$

Remark regarding the adjoint equation: The γ^ν are to be Hermitian; that is,

$$\gamma^{\nu *} = \gamma^{\nu T} \, , \qquad (\gamma^{\nu T})_{\alpha\beta} = (\gamma^\nu)_{\beta\alpha} \, . \qquad [3.13]$$

Here, the Hermiticity of the γ^ν is to be valid in the coordinates x^1, x^2, x^3, x^4, that is, with the imaginary time coordinate. Since not all of the four coordinates are real, the sign of the term with x_4 must be changed when forming the complex conjugate of Eq. [3.7]:

$$-\frac{\partial \psi^*}{\partial x^4}\gamma^4 + \sum_{k=1}^{3} \frac{\partial \psi^*}{\partial x^k}\gamma^k + m\psi^* = 0 \, . \qquad [3.14]$$

If we multiply with γ^4 from the right, then, because $\gamma^4\gamma^k = -\gamma^k\gamma^4$ and with

$$\psi^*\gamma^4 \equiv \bar{\psi}\,, \tag{3.15}$$

we get from [3.14]

$$\bar{\psi}\left(\gamma^\nu \frac{\overleftarrow{\partial}}{\partial x^\nu} - m\right) = 0\,.$$

Furthermore,

$$j^4 = C\,(\bar{\psi}\gamma^4\psi) = C\,(\psi^*\psi) = ij^0\,.$$

Since j^0 is to be the electric charge density, we set $C = ie$. Thus,

$$j^\nu = ie(\bar{\psi}\gamma^\nu\psi)\,. \tag{3.16}$$

We will now quantize the amplitudes and, indeed, since we know empirically that electrons satisfy the exclusion principle (there is also a theoretical basis for this [2]), we will follow the scheme given at the end of Section 2. This invention (by Jordan and Wigner [3]) is very useful although its physical meaning appears to be obscure: the sign of an expression in the amplitudes becomes dependent upon the numbering of the normal modes.

We therefore expand

$$\left.\begin{aligned} \psi_\varrho &= \sum_r A_r u_\varrho^{(r)}(x) \\ \psi_\varrho^* &= \sum_r A_r^\dagger u_\varrho^{(r)*}(x) \\ \bar{\psi}_\varrho &= \sum_r A_r^\dagger \bar{u}_\varrho^{(r)}(x) \end{aligned}\right\}, \tag{3.17}$$

and for the quantization we demand that

$$\left.\begin{aligned} \{A_r, A_s^\dagger\} &\equiv A_r A_s^\dagger + A_s^\dagger A_r = \delta_{rs} \\ \{A_r, A_s\} &= \{A_r^\dagger, A_s^\dagger\} = 0 \end{aligned}\right\}. \tag{3.18}$$

[2] W. Pauli, *Rev. Mod. Phys.* **13**, 203 (1941).
[3] P. Jordan and E. P. Wigner, *Z. Physik* **45**, 751 (1928).

Because of the completeness of the system,

$$\int j^0 \mathrm{d}^3x = e \sum_r A_r^\dagger A_r \, . \qquad [3.19]$$

If we now interpret $A_r^\dagger A_r$ as N_r, the number of particles in the state $u^{(r)}$, then $\int j^0 \, \mathrm{d}^3x/e$ is always positive. We therefore obtain a theory with only positive numbers of particles. By dividing the states into those with positive and those with negative frequencies, we can also construct a theory which describes electrons and positrons (see Sec. 4).

We first proceed to a generalization of the completeness relation. In nonrelativistic form this relation is

$$\sum_r u_\alpha^{(r)}(\boldsymbol{x}, t)\, u_\beta^{(r)*}(\boldsymbol{x}', t) = \delta_{\alpha\beta}\, \delta^3(\boldsymbol{x} - \boldsymbol{x}') \, . \qquad [3.20]$$

We can, however, free ourselves from the assumption of equal times. If we set $(\boldsymbol{x}, t) \equiv x$ and multiply Eq. [3.20] by γ^4, then we can write

$$\sum_r u_\alpha^{(r)}(x)\, \overline{u}_\beta^{(r)}(x') = - iS_{\alpha\beta}(x - x') \, . \qquad [3.21]$$

Here S is determined by the properties

$$S_{\alpha\beta}(\boldsymbol{x} - \boldsymbol{x}', 0) = i(\gamma^4)_{\alpha\beta}\, \delta^3(\boldsymbol{x} - \boldsymbol{x}') \, , \qquad [3.22]$$

$$\left(\gamma \frac{\partial}{\partial x} + m\right) S = 0 \, , \qquad S \left(\gamma \frac{\overleftarrow{\partial}}{\partial x'} - m\right) = 0 \, . \qquad [3.23]$$

That is, S is that solution of the Dirac equation which goes over into $i\gamma^4 \delta^3(\boldsymbol{x} - \boldsymbol{x}')$ for $t = 0$. This suffices to determine S uniquely, since the system of the Dirac equations is of first order. The determination of S can be reduced to the solution of a differential equation of second order. Because

$$\left(\gamma \frac{\partial}{\partial x} + m\right)\left(\gamma \frac{\partial}{\partial x} - m\right) \equiv \Box - m^2 \, ,$$

and if $\Delta(x)$ is defined by

$$\left.\begin{aligned} &(\Box - m^2)\Delta(x) = 0 \\ &\Delta(\boldsymbol{x}, 0) = 0 \\ &\left(\frac{\partial \Delta}{\partial t}\right)_{\boldsymbol{x},0} = -\delta^3(\boldsymbol{x}) \end{aligned}\right\}, \qquad [3.24]$$

then

$$S(x) = \left(\gamma \frac{\partial}{\partial x} - m\right)\Delta(x)\,. \qquad [3.25]$$

From the definition of $S(x)$ in Eq. [3.21] and from Eqs. [3.17], [3.18] it immediately follows that

$$\{\psi_\alpha(x), \bar{\psi}_\beta(x')\} = -iS_{\alpha\beta}(x - x')\,. \qquad [3.26]$$

4. SIGN OF THE ENERGY; HOLE THEORY

The quantity

$$E_{rr} = i\int u^{(r)*} \frac{\partial u^{(r)}}{\partial t}\,\mathrm{d}^3x \qquad [4.1]$$

is constant in time for the same reason as is the charge. That is,

$$E_{rs} = -\int \bar{u}^{(r)} \gamma^4 \frac{\partial u^{(s)}}{\partial x^4}\,\mathrm{d}^3x = \int \bar{u}^{(r)} \left(\boldsymbol{\gamma}\cdot\frac{\partial}{\partial \boldsymbol{x}} + m\right) u^{(s)}\,\mathrm{d}^3x\,,$$

the diagonal elements of which are constant. We now choose the $u^{(r)}$ so that E_{rs} is diagonal:

$$E_{rs} = \delta_{rs}\cdot\omega_r\cdot\varepsilon_r\,; \qquad \omega_r > 0\,, \qquad \varepsilon_r = \pm 1\,. \qquad [4.2]$$

For every solution with $\varepsilon_r > 0$ there exists one with $\varepsilon_r < 0$. The ε_r classifies the states according to positive and negative energy. For example, for plane waves,

$$\varepsilon_r = +1: \qquad u^{(r)} = C^{(r)}\cdot\frac{1}{\sqrt{G}}\exp\left[i(\boldsymbol{k}_r\cdot\boldsymbol{x} - \omega_r t)\right],$$

$$\varepsilon_r = -1: \qquad u^{(r)} = C^{(r)}\cdot\frac{1}{\sqrt{G}}\exp\left[i(\boldsymbol{k}_r\cdot\boldsymbol{x} + \omega_r t)\right].$$

Quite generally, any regular function of time $\psi(t)$ (which vanishes sufficiently fast at infinity) can be split into a part with positive and a part with negative frequencies by employing Fourier decomposition. This can also be accomplished without a Fourier decomposition as follows. One defines

$$\left.\begin{aligned} \psi^+(t) &= \frac{1}{2\pi i}\int_{C_+} \psi(t-\varepsilon\tau)\frac{\mathrm{d}\tau}{\tau} \\ \psi^-(t) &= -\frac{1}{2\pi i}\int_{C_-} \psi(t-\varepsilon\tau)\frac{\mathrm{d}\tau}{\tau} \end{aligned} \quad (\varepsilon>0) \right\} . \tag{4.3}$$

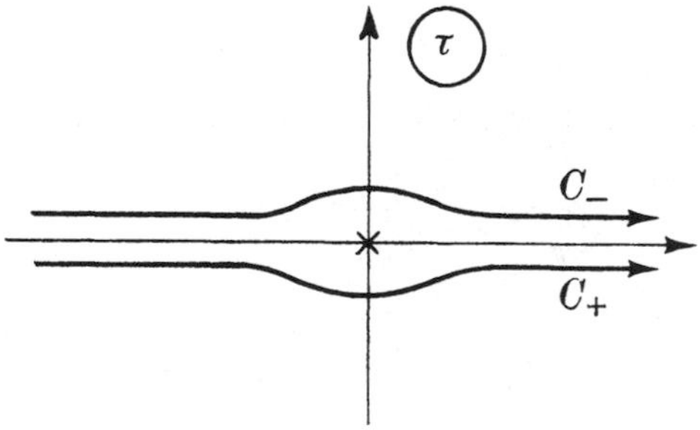

Figure 4.1

This is based upon the following:

$$\frac{1}{2\pi i}\int_{C_+} \exp[-i\omega(t-\varepsilon\tau)]\frac{\mathrm{d}\tau}{\tau} = \frac{\exp[-i\omega t]}{2\pi i}\int_{C_+} \exp[i\omega\varepsilon\tau]\frac{\mathrm{d}\tau}{\tau} .$$

Evaluation using residues yields:

for $\omega>0$, the path must be closed in the upper half-plane to ensure that $\exp[i\omega\varepsilon\tau]$ remains bounded for large τ;

for $\omega<0$, the path must be closed in the lower half-plane.

Thus,

$$\frac{1}{2\pi i}\int_{C_+} \exp[-i\omega(t-\varepsilon\tau)]\frac{\mathrm{d}\tau}{\tau} = \begin{cases} \exp[-i\omega t] & (\omega>0) \\ 0 & (\omega<0) \end{cases} .$$

Then,

$$\psi^+(t) + \psi^-(t) = \psi(t) . \tag{4.4}$$

In addition,

$$i\big(\psi^+(t) - \psi^-(t)\big) \equiv \psi^1(t) = \frac{1}{\pi}\mathscr{P}\int_{-\infty}^{+\infty} \psi(t-\varepsilon\tau)\,\frac{d\tau}{\tau}\,, \qquad [4.5]$$

where $\mathscr{P}$ is the principal value defined on the real axis,

$$\mathscr{P}\int_{-\infty}^{+\infty} f(t)\,dt = \lim_{\varepsilon\to 0}\left[\int_{-\infty}^{-\varepsilon} f(t)\,dt + \int_{+\varepsilon}^{+\infty} f(t)\,dt\right]. \qquad [4.6]$$

With this, therefore,

$$\left.\begin{aligned}\psi^+ &= \tfrac{1}{2}(\psi - i\psi^1)\\ \psi^- &= \tfrac{1}{2}(\psi + i\psi^1)\end{aligned}\right\}. \qquad [4.7]$$

Correspondingly, one can make the separation

$$\Delta = \Delta^+ + \Delta^-, \qquad \Delta^\pm = \tfrac{1}{2}(\Delta \mp i\Delta^1)\,; \qquad [4.8]$$

$$S = S^+ + S^-, \qquad S^\pm = \tfrac{1}{2}(S \mp iS^1)\,. \qquad [4.9]$$

Hole theory is symmetric with respect to positively and negatively charged particles. It is obtained by setting

$$\left.\begin{aligned}A_r^* A_r &= N_r\\ A_r A_r^* &= 1 - N_r\end{aligned}\right\}(\varepsilon_r > 0)\,, \qquad \left.\begin{aligned}A_r A_r^* &= N_r\\ A_r^* A_r &= 1 - N_r\end{aligned}\right\}(\varepsilon_r < 0)\,; \qquad [4.10]$$

that is,

$$\left.\begin{aligned}\langle A_r^* A_r\rangle_0 &= 0\,, \quad \langle A_r A_r^*\rangle_0 = 1\,, \quad (\varepsilon_r > 0)\\ \langle A_r^* A_r\rangle_0 &= 1\,, \quad \langle A_r A_r^*\rangle_0 = 0\,, \quad (\varepsilon_r < 0)\end{aligned}\right\}. \qquad [4.11]$$

Here, $\langle\,\rangle_0$ represents the expectation value with respect to the vacuum, which is defined as the state of lowest energy. That the state so defined is really the vacuum can be shown as follows. If E denotes the energy, then

$$E = i\int \psi^* \frac{\partial\psi}{\partial t}\,d^3x = \sum_r \omega_r \varepsilon_r A_r^* A_r = \sum_{r,\varepsilon_r>0} \omega_r N_r + \sum_{r,\varepsilon_r<0} \omega_r N_r - \sum_{r,\varepsilon_r<0} \omega_r\,.$$

Therefore

$$E = \sum_r \omega_r N_r - \sum_{r,\varepsilon_r<0} \omega_r\,. \qquad [4.12]$$

The second term is a (divergent) constant. The minimum of the energy occurs at $N_r = 0$. That is, the state of lowest energy and the state without particles are the same.

With Eq. [4.11] one easily finds that

$$\left.\begin{aligned} \langle \psi_\alpha(x)\bar{\psi}_\beta(x')\rangle_0 &= -\,iS^+_{\alpha\beta}(x-x') = \tfrac{1}{2}(-\,iS - S^1)_{\alpha\beta} \\ \langle \bar{\psi}_\beta(x')\psi_\alpha(x)\rangle_0 &= -\,iS^-_{\alpha\beta}(x-x') = \tfrac{1}{2}(S^1 - iS)_{\alpha\beta} \end{aligned}\right\} . \qquad [4.13]$$

From this we obtain

$$\langle [\psi_\alpha(x), \bar{\psi}_\beta(x')]\rangle_0 = -\,S^1_{\alpha\beta}(x-x') . \qquad [4.14]$$

5. CONSTRUCTION OF THE INVARIANT FUNCTIONS

In this section the functions Δ, Δ^1, etc., defined only implicitly until now, will be explicitly constructed. We have

$$\Delta(\boldsymbol{x}, t) = -\frac{i}{2\cdot(2\pi)^3}\int\Big(\exp[i(\boldsymbol{k}\cdot\boldsymbol{x} - \omega t)] - \exp[-\,i(\boldsymbol{k}\cdot\boldsymbol{x} - \omega t)]\Big)\frac{\mathrm{d}^3k}{\omega} , \qquad [5.1]$$

where $\omega = +\sqrt{m^2 + k^2}$.

Proof:

$$\Delta(x) = -\,\Delta(-\,x) , \qquad (x) \equiv (\boldsymbol{x}, t) ,$$

$$\Delta(\boldsymbol{x}, t) = +\Delta(-\,\boldsymbol{x}, t) ,$$

$$\Delta(\boldsymbol{x}, t) = -\,\Delta(\boldsymbol{x}, -\,t) .$$

Thus,

$$\frac{\partial\Delta}{\partial t} = -\frac{i}{2\cdot(2\pi)^3}\int(-\,i\omega)\Big(\exp[i(kx)] + \exp[-\,i(kx)]\Big)\frac{\mathrm{d}^3k}{\omega} ,$$

$$(kx) \equiv \boldsymbol{k}\cdot\boldsymbol{x} - \omega t ,$$

$$\left.\frac{\partial\Delta}{\partial t}\right|_{t=0} = -\frac{1}{2}\left(\frac{1}{2\pi}\right)^3\cdot 2\int\exp[i\boldsymbol{k}\cdot\boldsymbol{x}]\,\mathrm{d}^3k = -\,\delta^3(\boldsymbol{x}) .$$

Therefore, Δ possesses all the required properties. Other representations are the following:

$$\Delta(\boldsymbol{x}, t) = -\left(\frac{1}{2\pi}\right)^3\int\exp[i\boldsymbol{k}\cdot\boldsymbol{x}]\sin\omega t\,\frac{\mathrm{d}^3k}{\omega} \qquad [5.2]$$

and

$$\Delta(x) = -\frac{i}{(2\pi)^3}\int \exp[i(kx)]\,\varepsilon(k)\,\delta(k^2+m^2)\,\mathrm{d}^4k\,, \qquad k^2 \equiv (kk)\,, \tag{5.3}$$

where

$$\varepsilon(k) = \begin{cases} +1 & k_0 > 0\,, \\ -1 & k_0 < 0\,. \end{cases}$$

Proof of Eq. [5.3]:

$$\int F(\boldsymbol{k}, k_0)\,\delta(k^2+m^2)\,\mathrm{d}k_0 = \int \delta(k_0^2-\omega^2)\,F(\boldsymbol{k}, k_0)\,\mathrm{d}k_0\,.$$

If $f(z_0) = 0$, then

$$\delta\big(f(z)\big) = \sum_{z_0} \frac{\delta(z-z_0)}{|f'(z_0)|}\,.$$

Thus,

$$\int \delta(k_0^2-\omega^2)\,F(\boldsymbol{k}, k_0)\,\mathrm{d}k_0 = \frac{1}{2\omega}\,[F(\boldsymbol{k}, \omega) + F(\boldsymbol{k}, -\omega)],$$

and this reduces Eq. [5.3] to Eq. [5.1].

If we separate positive and negative frequencies,

$$\left.\begin{aligned} \Delta^+(x) &= -\frac{i}{(2\pi)^3}\int \exp[i(\boldsymbol{k}\cdot\boldsymbol{x}-\omega t)]\,\frac{\mathrm{d}^3k}{2\omega} \\ \Delta^-(x) &= \frac{i}{(2\pi)^3}\int \exp[i(\boldsymbol{k}\cdot\boldsymbol{x}+\omega t)]\,\frac{\mathrm{d}^3k}{2\omega} \end{aligned}\right\}, \tag{5.4}$$

we obtain

$$\left.\begin{aligned} \Delta^1(x) \equiv i(\Delta^+-\Delta^-) &= \left(\frac{1}{2\pi}\right)^3 \int \Big\{\big(\exp[i(\boldsymbol{k}\cdot\boldsymbol{x}-\omega t)] \\ &\qquad + \exp[-i(\boldsymbol{k}\cdot\boldsymbol{x}-\omega t)]\big)\Big\}\,\frac{\mathrm{d}^3k}{2\omega} \\ &= \left(\frac{1}{2\pi}\right)^3 \int \exp[i\boldsymbol{k}\cdot\boldsymbol{x}]\cos\omega t\,\frac{\mathrm{d}^3k}{\omega} \\ &= \left(\frac{1}{2\pi}\right)^3 \int \exp[i(kx)]\,\delta(k^2+m^2)\,\mathrm{d}^4k \end{aligned}\right\}. \tag{5.5}$$

Note: In Eq. [5.3], although $\varepsilon(k)$ appears to disturb the relativistic invariance, $\varepsilon(k)\delta(k^2+m^2)$ is again Lorentz invariant, at least with respect to Lorentz transformations without time reversal $(k_0 \to -k_0)$.

We define another solution of the inhomogeneous wave equation:

$$(\Box - m^2)\bar{\Delta}(x) = -\delta^4(x)\,. \qquad [5.6]$$

This does not yet fix $\bar{\Delta}$ uniquely. We fix $\bar{\Delta}$ by the relation

$$\bar{\Delta}(x) = -\tfrac{1}{2}\varepsilon(x)\Delta(x)\,; \qquad \varepsilon(x) = \begin{cases} +1 & t>0 \\ -1 & t<0 \end{cases}. \qquad [5.7]$$

This is, in fact, a solution of Eq. [5.6]:

$$\Box\bar{\Delta} = -\frac{1}{2}\varepsilon\Box\Delta - \frac{\partial\varepsilon}{\partial x^\mu}\frac{\partial\Delta}{\partial x^\mu} - \frac{1}{2}(\Box\varepsilon)\Delta\,,$$

$$\left.\begin{aligned} &\frac{\partial\varepsilon}{\partial x^i} = 0 \quad (i \neq 4), \quad \frac{\partial\varepsilon}{\partial x^4} = -i\frac{\partial\varepsilon}{\partial t} = -2i\delta(t) \\ &\left(\frac{\partial\Delta}{\partial x^4}\right)_{t=0} = -i\left(\frac{\partial\Delta}{\partial t}\right)_{t=0} = +i\delta^3(\boldsymbol{x}) \end{aligned}\right\} \frac{\partial\varepsilon}{\partial x^\mu}\frac{\partial\Delta}{\partial x^\mu} = +2\delta^4(x),$$

$$(\Box\varepsilon)\Delta = -\frac{\partial^2\varepsilon}{\partial t^2}\Delta = -2\delta'(t)\Delta = +2\delta(t)\frac{\partial\Delta}{\partial t} = -2\delta^4(x)\,;$$

this results in

$$(\Box - m^2)\bar{\Delta} = -\delta^4(x)\,. \qquad \text{Q.E.D.}$$

Note: It is true that

$$\bar{\Delta}(-x) = +\bar{\Delta}(x)\,. \qquad [5.8]$$

Furthermore,

$$\Delta(x) = 0 \quad \text{implies} \quad \bar{\Delta}(x) = 0 \qquad \text{for } |\boldsymbol{x}|^2 > t^2,$$

since

$$\Delta(\boldsymbol{x}, 0) = 0\,, \qquad \left(\frac{\partial\Delta}{\partial t}\right)_{t=0} = -\delta^3(\boldsymbol{x})\,,$$

and since Δ is an invariant function of $(x, x) = \boldsymbol{x}^2 - t^2 = -\lambda$. (On the contrary, Δ^1 is, in general, not equal to zero for $\boldsymbol{x}^2 > t^2$.)

Here $\bar{\Delta}$ is uniquely determined by the properties

1. $$(\Box - m^2)\bar{\Delta}(x) = -\delta^4(x)\,;$$

2. $$\bar{\Delta}(x) = 0 \qquad (\boldsymbol{x}^2 > t^2)\,;$$

3. $$\bar{\Delta}(x) = \bar{\Delta}(-x)\,.$$

This is true because of the following reasons: $\bar{\Delta}$ is determined by property 1 to within solutions of the homogeneous differential equation. However, additional terms proportional to Δ^1 are excluded by property 2, while terms proportional to Δ are excluded by property 3.

Additional solutions of Eq. [5.6] are the advanced and retarded Δ-functions:

Retarded Δ-function:

$$\Delta^{\text{ret}}(x) \equiv \bar{\Delta} - \tfrac{1}{2}\Delta = -\tfrac{1}{2}(1+\varepsilon)\Delta\,;$$

Advanced Δ-function:

$$\Delta^{\text{adv}}(x) \equiv \bar{\Delta} + \tfrac{1}{2}\Delta = +\tfrac{1}{2}(1-\varepsilon)\Delta = \Delta^{\text{ret}}(-x)\,.$$

We have

$$\left.\begin{aligned} \Delta^{\text{ret}}(x) &= \begin{cases} -\Delta(x) & t>0 \\ 0 & t<0 \end{cases} \\ \Delta^{\text{adv}}(x) &= \begin{cases} 0 & t>0 \\ +\Delta(x) & t<0 \end{cases} \end{aligned}\right\}. \qquad [5.9]$$

For the construction of these functions see Schwinger[4] and Section 13 of this volume.

These functions serve to solve the differential equation

$$(\Box - m^2)\varphi(x) = -f(x)\,. \qquad [5.10]$$

[4] J. Schwinger, *Phys. Rev.* **74**, 1439 (1948); **75**, 651 (1949); **76**, 790 (1949).

That is,

$$\varphi_{\text{ret}}(x) = \int \Delta^{\text{ret}}(x-x')f(x')\,\mathrm{d}^4x' = -\int_{t'<t} \Delta(x-x')f(x')\,\mathrm{d}^4x',$$

$$\varphi_{\text{adv}}(x) = \int \Delta^{\text{adv}}(x-x')f(x')\,\mathrm{d}^4x' = +\int_{t'>t} \Delta(x-x')f(x')\,\mathrm{d}^4x',$$

$$\bar{\varphi}(x) = \frac{1}{2}(\varphi_{\text{ret}} + \varphi_{\text{adv}}) = \int f(x')\bar{\Delta}(x-x')\,\mathrm{d}^4x'.$$

We have

$$\left.\begin{aligned} \bar{\Delta}(x) &= \tfrac{1}{2}\big(\Delta^{\text{ret}}(x) + \Delta^{\text{adv}}(x)\big) \\ \Delta(x) &= \Delta^{\text{adv}}(x) - \Delta^{\text{ret}}(x) \end{aligned}\right\}. \qquad [5.11]$$

Without proof, let it be stated that

$$\bar{\Delta}(x) = \begin{cases} \dfrac{1}{4\pi}\,\delta(\lambda) - \dfrac{m^2}{8\pi} J_1(m\sqrt{\lambda})\,\dfrac{1}{m\sqrt{\lambda}}, & \lambda = -(xx) \geqslant 0, \\ 0, & \lambda < 0. \end{cases} \qquad [5.12]$$

For $m=0$, $\bar{\Delta}$ is different from zero only on the light cone.

A representation for $\bar{\Delta}$ in momentum space is

$$\bar{\Delta}(x) = \left(\frac{1}{2\pi}\right)^4 \mathscr{P}\int \frac{\exp[i(kx)]}{k^2+m^2}\,\mathrm{d}^4k.$$

Here, $\mathscr{P}$ represents the principal value with respect to k_0. It is easily verified that Eq. [5.6] is satisfied.

For a first-order pole, the principal value can be defined in the complex plane as follows (Fig. 5.1):

$$\mathscr{P}\int_{-\infty}^{+\infty} \frac{f(x)\,\mathrm{d}x}{x-a} = \frac{1}{2}\left(\int_{C_+} \frac{f(z)\,\mathrm{d}z}{z-a} + \int_{C_-} \frac{f(z)\,\mathrm{d}z}{z-a}\right).$$

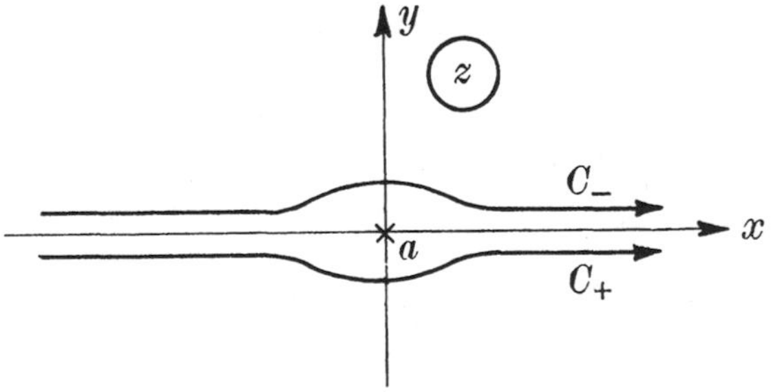

Figure 5.1

Now, one can form

$$\left.\begin{aligned} \Delta^C(x) &\equiv \Delta^1 - 2i\bar{\Delta} \\ &= \frac{-2i}{(2\pi)^4}\mathscr{P}\int \exp[i(kx)]\left[\frac{1}{k^2+m^2} + i\pi\delta(k^2+m^2)\right] \mathrm{d}^4k \\ &= \frac{-2i}{(2\pi)^4}\int_C \frac{\exp[i(kx)]}{k^2+m^2}\,\mathrm{d}^4k \end{aligned}\right\} \quad [5.13]$$

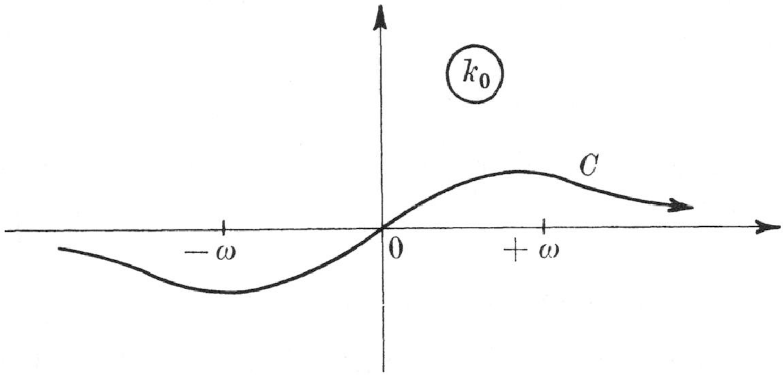

Figure 5.2

where the path C is defined in Figure 5.2. Following Heisenberg,[5] we define

$$\frac{1}{2\pi i}\left(\frac{1}{k^2+m^2} + i\pi\delta(k^2+m^2)\right) \equiv \delta_+(k^2+m^2)\,.$$

Furthermore,

$$\left.\begin{aligned} \Delta^C &= 2i\Delta^+ \quad (t>0)\,, \quad \text{outgoing waves} \\ \Delta^C &= -2i\Delta^- \quad (t<0)\,, \quad \text{incoming waves} \end{aligned}\right\}. \quad [5.14]$$

This function appears in the following context. We had Eq. [3.26] and Eq. [4.14],

$$\{\psi_\alpha(x), \bar{\psi}_\beta(x')\} = -iS_{\alpha\beta}(x-x')\,,$$

$$\langle[\psi_\alpha(x), \bar{\psi}_\beta(x')]\rangle_0 = -S^1_{\alpha\beta}(x-x')\,.$$

[5] W. HEISENBERG, *Z. Physik* **120**, 513 (1943). See also P. A. M. DIRAC *Quantum Mechanics*, 2nd edition (Oxford: Clarendon Press, 1935), p. 200.

If, following Dyson,[6] one defines the chronological or time-ordered product,

$$\left.\begin{aligned} P\big(A(x)B(x')\big) &= A(x)B(x') \qquad \text{for } t > t' \\ &= B(x')A(x) \qquad \text{for } t < t' \end{aligned}\right\}, \qquad [5.15]$$

or

$$P\big(A(x)B(x')\big) = \tfrac{1}{2}\{A(x), B(x')\} + \tfrac{1}{2}\,\varepsilon(x - x')[A(x), B(x')] \quad [5.16]$$

(this is not an invariant definition), then it immediately follows that

$$\langle P\big(\psi_\alpha(x), \bar\psi_\beta(x')\big)\rangle_0\, \varepsilon(x - x') = -\tfrac{1}{2} S^C_{\alpha\beta}(x - x') \,. \qquad [5.17]$$

For the physical significance of Δ^C, see Section 28 (for $m = 0$).

6. CHARGE-CONJUGATED QUANTITIES

We return to the Dirac equations, [3.7], [3.10], [3.16]:

$$\left(\gamma \frac{\partial}{\partial x} + m\right)\psi = 0\,, \qquad \bar\psi\left(\gamma \frac{\overleftarrow{\partial}}{\partial x} - m\right) = 0\,,$$

$$j^\mu = ie(\bar\psi\gamma^\mu\psi)\,.$$

From $\bar\psi$ a new, so-called charge-conjugated, solution of Eq. [3.7] is obtained by forming

$$\left.\begin{aligned} \psi' &= C\bar\psi \\ \bar\psi' &= C^{-1}\psi \end{aligned}\right\} \qquad [6.1]$$

if

$$\gamma^{\mu T} = -\,C^{-1}\gamma^\mu C \qquad [6.2]$$

is valid for C, where the transposed matrices $\gamma^{\mu T}$ are defined in [3.13].

Such a C exists since there is always a similarity transformation which leads from one matrix system to another if both realize the same algebra.

[6] F. J. Dyson, *Phys. Rev.* **75**, 486 (1949).

Because of the Hermiticity,

$$(\gamma^\mu)^* = (\gamma^\mu)^T ,$$

we can prove that

$$\left(\gamma \frac{\partial}{\partial x} + m\right) \psi' = 0 .$$

Since

$$\bar{\psi} \left(\gamma \frac{\overleftarrow{\partial}}{\partial x} - m\right) = 0$$

is the same as

$$\left(\gamma^T \frac{\partial}{\partial x} - m\right) \bar{\psi} = 0 ,$$

then [6.2] implies

$$\left(- C^{-1}\gamma C \frac{\partial}{\partial x} - m\right) \bar{\psi} = 0 , \quad \left(\gamma \frac{\partial}{\partial x} + m\right) C\bar{\psi} = 0 . \quad \text{Q.E.D.}$$

One easily finds that

$$\left.\begin{aligned} CC^\dagger &= 1 \\ C^T &= - C \end{aligned}\right\} . \qquad [6.3]$$

In the presence of an external field, the Dirac equations read

$$\left(\gamma^\nu \left(\frac{\partial}{\partial x^\nu} - ie\mathscr{A}_\nu\right) + m\right) \psi = 0 ,$$

$$\bar{\psi} \left(\gamma^\nu \left(\frac{\overleftarrow{\partial}}{\partial x^\nu} + ie\mathscr{A}_\nu\right) - m\right) = 0 .$$

Thus,

$$\left(\gamma^{\nu T} \left(\frac{\partial}{\partial x^\nu} + ie\mathscr{A}_\nu\right) - m\right) \bar{\psi} = 0 ,$$

and with [6.2]

$$\left(\gamma^\nu \left(\frac{\partial}{\partial x^\nu} + ie\mathscr{A}_\nu\right) + m\right) C\bar{\psi} = 0 ,$$

$$\left(\gamma^\nu \left(\frac{\partial}{\partial x^\nu} + ie\mathscr{A}_\nu\right) + m\right) \psi' = 0 ;$$

that is, ψ' satisfies the same Dirac equation as ψ except that the sign of the charge in the external field is reversed. In the field-free case one need not distinguish between ψ and ψ'.

In a c-number theory, the current is given by

$$j^\mu(x) = ie(\bar{\psi}\gamma^\mu\psi) .$$

We form

$$j'^\mu(x) = ie(\bar{\psi}'\gamma^\mu\psi') , \qquad \psi' = C\bar{\psi} , \qquad \bar{\psi}' = -\psi C^{-1} ,$$

so that

$$j'^\mu(x) = -ie(\psi C^{-1}\gamma^\mu C\bar{\psi}) = ie(\psi\gamma^{\mu T}\bar{\psi}) .$$

In a c-number theory, therefore,

$$j'^\mu(x) = j^\mu(x) ,$$

which is unsatisfying.

In a q-number theory, because of the quantization according to the exclusion principle, we can do something more satisfying. Following Heisenberg, we set

$$j^\mu = \tfrac{1}{2} ie(\bar{\psi}\gamma^\mu\psi - \psi\gamma^{\mu T}\bar{\psi}) = -\tfrac{1}{2} ie[\psi_\alpha, \bar{\psi}_\beta]\gamma^\mu_{\beta\alpha} . \qquad [6.4]$$

Quite generally, for any quantity, one should write

$$\bar{\psi}F\psi \rightarrow \tfrac{1}{2}(\bar{\psi}F\psi - \psi F^T\bar{\psi}) = -\tfrac{1}{2}[\psi_\alpha, \bar{\psi}_\beta]F_{\beta\alpha} . \qquad [6.5]$$

Then,

$$j'^\mu(x) = \frac{ie}{2}(\bar{\psi}'\gamma^\mu\psi' - \psi'\gamma^{\mu T}\bar{\psi}') = -j^\mu(x) . \qquad [6.6]$$

If one expands,

$$\psi_\alpha(x) = \sum_r A_r u_\alpha^{(r)}(x) ,$$

where

$$\{A_r, A_s^*\} = \delta_{rs} , \qquad \{A_r, A_s\} = \{A_r^*, A_s^*\} = 0 ,$$
$$\langle A_r A_s^*\rangle_0 = \delta_{rs}(1+\varepsilon_r) ,$$
$$\langle A_r^* A_s\rangle_0 = \delta_{rs}(1-\varepsilon_r) ,$$

then the total charge is

$$e=\int j^0\,\mathrm{d}^3x=\tfrac{1}{2}\sum_r [A_r^*, A_r]=\sum_r (N_r^+-\tfrac{1}{2})-\sum_r (N_r^- -\tfrac{1}{2})\,. \quad [6.7]$$

The vacuum expectation value of the commutator in Eq. [6.4] is not well defined because of the singularity of S^1. If, by definition, one sets

$$[\psi_\alpha(x), \bar{\psi}_\beta(x)]=\lim_{x'\to x}\tfrac{1}{2}\{[\psi_\alpha(x), \bar{\psi}_\beta(x')]+[\psi_\alpha(x'), \bar{\psi}_\beta(x)]\}\,, \quad [6.8]$$

then

$$\langle j^\mu\rangle_0=0\,, \quad [6.9]$$

as it must be.

In addition, it is true that

$$\left[C^{-1}\begin{pmatrix}S^1(-x)\\ \bar{S}\ (-x)\\ S^C(-x)\end{pmatrix}C\right]^T=\begin{pmatrix}S^1(x)\\ \bar{S}\ (x)\\ S^C(x)\end{pmatrix}; \quad [6.10]$$

that is

$$\left.\begin{aligned}C^{-1}_{\beta\varrho}S^1_{\varrho\sigma}(-x)\,C_{\sigma\alpha}&=\ \ S^1_{\alpha\beta}(x)\\ C^{-1}_{\beta\varrho}S^\pm_{\varrho\sigma}(-x)\,C_{\sigma\alpha}&=-\,S^\mp_{\alpha\beta}(x)\\ C^{-1}_{\beta\varrho}S_{\varrho\sigma}(-x)\,C_{\sigma\alpha}&=-\,S_{\alpha\beta}(x)\end{aligned}\right\}. \quad [6.11]$$

Furthermore the above imply

$$\left.\begin{aligned}\{\psi'_\alpha(x), \bar{\psi}'_\beta(x')\}&=\{\psi_\alpha(x), \bar{\psi}_\beta(x')\}\\ \langle[\psi'_\alpha(x), \bar{\psi}'_\beta(x')]\rangle_0&=\langle[\psi_\alpha(x), \bar{\psi}_\beta(x')]\rangle_0\end{aligned}\right\}. \quad [6.12]$$

Chapter 2. Response to an External Field: Charge Renormalization

7. VACUUM EXPECTATION VALUES OF EXPRESSIONS BILINEAR IN THE CURRENT

We have

$$j^\mu(x) = ie(\bar{\psi}(x)\gamma^\mu\psi(x)) ,$$

or, more exactly,

$$j^\mu(x) = -\frac{ie}{2}[\psi_\alpha(x), \bar{\psi}_\beta(x)]\gamma^\mu_{\beta\alpha} .$$

Note: This rearrangement has only the effect that $\langle j^\mu(x)\rangle_0 = 0$, and has, therefore, no influence on our calculation.

We wish to calculate the following expectation value:

$$\begin{aligned}\langle j^\mu(x)\, j^\nu(x')\rangle_0 &= -\, e^2\langle\bar{\psi}_\alpha(x)\psi_\beta(x)\bar{\psi}_\varrho(x')\,\psi_\sigma(x')\rangle_0\gamma^\mu_{\alpha\beta}\gamma^\nu_{\varrho\sigma} \\ &= -\, e^2\langle\bar{\psi}_\alpha(x)\psi_\sigma(x')\rangle_0\langle\psi_\beta(x)\,\bar{\psi}_\varrho(x')\rangle_0\gamma^\mu_{\alpha\beta}\gamma^\nu_{\varrho\sigma} .\end{aligned}$$

With

$$S^- = \tfrac{1}{2}(S + iS^1) ,$$

$$S^+ = \tfrac{1}{2}(S - iS^1) ,$$

$$\langle\psi_\alpha(x)\,\bar{\psi}_\beta(x')\rangle_0 = -\, iS^+_{\alpha\beta}(x - x') ,$$

$$\langle\bar{\psi}_\beta(x')\,\psi_\alpha(x)\rangle_0 = -\, iS^-_{\alpha\beta}(x - x') ,$$

we get

$$\begin{aligned}\langle j^\mu(x)\, j^\nu(x')\rangle_0 &= +\, e^2 S^-_{\sigma\alpha}(x' - x)\, S^+_{\beta\varrho}(x - x')\,\gamma^\mu_{\alpha\beta}\gamma^\nu_{\varrho\sigma} \\ &= +\, e^2\,\mathrm{Tr}\,\{\gamma^\mu S^+(x - x')\gamma^\nu S^-(x' - x)\} ,\end{aligned}$$

$$\langle j^\nu(x')\, j^\mu(x)\rangle_0 = +\, e^2\,\mathrm{Tr}\,\{\gamma^\mu S^-(x - x')\gamma^\nu S^+(x' - x)\} ,$$

and

$$\left.\begin{aligned}
&\tfrac{1}{2}\langle\{j^\mu(x),\, j^\nu(x')\}\rangle_0 \\
&\qquad = \frac{e^2}{4}\,\mathrm{Tr}\,\{\gamma^\mu S(x-x')\gamma^\nu S(x'-x) \\
&\qquad\qquad + \gamma^\mu S^1(x-x')\gamma^\nu S^1(x'-x)\} \\
&\tfrac{1}{2}\langle[j^\mu(x),\, j^\nu(x')]\rangle_0 \\
&\qquad = \frac{ie^2}{4}\,\mathrm{Tr}\,\{\gamma^\mu S(x-x')\gamma^\nu S^1(x'-x) \\
&\qquad\qquad - \gamma^\mu S^1(x-x')\gamma^\nu S(x'-x)\}
\end{aligned}\right\}. \qquad [7.1]$$

Upon employing the useful relation

$$\tfrac{1}{4}\,\mathrm{Tr}\,(\gamma^\mu\gamma^\nu\gamma^\varrho\gamma^\sigma) = \delta_{\mu\nu}\delta_{\varrho\sigma} - \delta_{\mu\varrho}\delta_{\nu\sigma} + \delta_{\mu\sigma}\delta_{\nu\sigma}\,, \qquad [7.2]$$

one obtains

$$\begin{aligned}
\tfrac{1}{2}\langle\{(j^\mu(x),\, j^\nu(x')\}\rangle_0 = e^2 \Bigg\{ & \frac{\partial\Delta(x-x')}{\partial x^\mu}\,\frac{\partial\Delta(x'-x)}{\partial x'^\nu} \\
& + \frac{\partial\Delta(x-x')}{\partial x^\nu}\,\frac{\partial\Delta(x'-x)}{\partial x'^\mu} - \delta_{\mu\nu}\left(\frac{\partial\Delta(x-x')}{\partial x^\alpha}\,\frac{\partial\Delta(x'-x)}{\partial x'^\alpha}\right. \\
& \left. - m^2\Delta(x-x')\Delta(x'-x)\right) + (\Delta\to\Delta^1)\Bigg\}.
\end{aligned}$$

We write

$$\tfrac{1}{2}\langle\{j^\mu(x),\, j^\nu(x')\}\rangle_0 \equiv e^2\hat{K}_{\mu\nu}(x-x')\,, \qquad [7.3]$$

where

$$\begin{aligned}
\hat{K}_{\mu\nu}(\xi) = {} & 2\,\frac{\partial\Delta}{\partial\xi^\mu}\,\frac{\partial\Delta}{\partial\xi^\nu} + \delta_{\mu\nu}\left|-\left(\frac{\partial\Delta}{\partial\xi}\right)^2 - m^2\Delta\Delta\right] \\
& - 2\,\frac{\partial\Delta^1}{\partial\xi^\mu}\,\frac{\partial\Delta^1}{\partial\xi^\nu} + \delta_{\mu\nu}\left[+\left(\frac{\partial\Delta^1}{\partial\xi}\right)^2 + m^2\Delta^1\Delta^1\right].
\end{aligned} \qquad [7.4]$$

Analogously,

$$\tfrac{1}{2}\langle[j^\mu(x),\, j^\nu(x')]\rangle_0 \equiv -\,ie^2K_{\mu\nu}(x-x')\,, \qquad [7.5]$$

with

$$K_{\mu\nu}(\xi) = 2\left\{\frac{\partial\Delta}{\partial\xi^\mu}\frac{\partial\Delta^1}{\partial\xi^\nu} + \frac{\partial\Delta^1}{\partial\xi^\mu}\frac{\partial\Delta}{\partial\xi^\nu} + \delta_{\mu\nu}\left[-\frac{\partial\Delta}{\partial\xi^\alpha}\frac{\partial\Delta^1}{\partial\xi^\alpha} - m^2\Delta\Delta^1\right]\right\}. \qquad [7.6]$$

Transition to Momentum Space

We set

$$\left.\begin{aligned} K_{\mu\nu}(x-x') &= \left(\frac{1}{2\pi}\right)^4 \int \exp\left[ip(x-x')\right] K_{\mu\nu}(p)\,\mathrm{d}^4p \\ K_{\mu\nu}(p) &= \int \exp\left[-ipx\right] K_{\mu\nu}(x)\,\mathrm{d}^4x \end{aligned}\right\}. \qquad [7.7]$$

Furthermore,

$$\Delta(x) = -\frac{i}{(2\pi)^3}\int \exp\left[ikx\right]\varepsilon(k)\,\delta(k^2+m^2)\,\mathrm{d}^4k\,,$$

$$\Delta^1(x) = +\left(\frac{1}{2\pi}\right)^3 \int \exp\left[ikx\right]\delta(k^2+m^2)\,\mathrm{d}^4k\,.$$

It then follows that

$$\hat{K}_{\mu\nu}(p) = -\left(\frac{1}{2\pi}\right)^2 \cdot \int \delta(k^2+m^2)\,\delta\big((k-p)^2+m^2\big)\left[1+\varepsilon(k)\,\varepsilon(p-k)\right] \cdot\left[-k_\mu(p-k)_\nu - k_\nu(p-k)_\mu - \delta_{\mu\nu}\big(-(p-k)_\lambda k_\lambda + m^2\big)\right]\mathrm{d}^4k, \qquad [7.8]$$

$$K_{\mu\nu}(p) = \frac{-i}{(2\pi)^2} \cdot \int \delta(k^2+m^2)\,\delta\big((k-p)^2+m^2\big)\left[\varepsilon(k)+\varepsilon(p-k)\right]\cdot\left[-k_\mu(p-k)_\nu - k_\nu(p-k)_\mu - \delta_{\mu\nu}\big(-(p-k)_\lambda k_\lambda + m^2\big)\right]\mathrm{d}^4k\,. \qquad [7.9]$$

Note: 1. If, in addition, we employ

$$\varepsilon(k)+\varepsilon(p-k) = \varepsilon(p)\left[1+\varepsilon(k)\,\varepsilon(p-k)\right], \qquad [7.10]$$

then we see that

$$K_{\mu\nu}(p) = i\varepsilon(p)\,\hat{K}_{\mu\nu}(p)\,. \qquad [7.11]$$

2. It is easily verified that

$$\frac{\partial K_{\mu\nu}}{\partial x^\nu} = 0\,, \qquad [7.12]$$

$$\frac{\partial \hat{K}_{\mu\nu}}{\partial x^\nu} = 0\,, \qquad [7.13]$$

by using the fact that Δ and Δ^1 satisfy the homogeneous differential equation

$$\frac{\partial \hat{K}_{\mu\nu}}{\partial x^\nu} = 2\frac{\partial^2 \Delta}{\partial x^\mu \partial x^\nu}\frac{\partial \Delta}{\partial x^\nu} + 2\frac{\partial \Delta}{\partial x^\mu}\Box\Delta - 2\frac{\partial \Delta}{\partial x^\varrho}\frac{\partial^2 \Delta}{\partial x^\varrho \partial x^\mu} - 2m^2\Delta\frac{\partial \Delta}{\partial x^\mu}$$

$$-\,(\text{the same expression but with } \Delta \text{ replaced by } \Delta^1)$$

$$= 2(\Box - m^2)\Delta\cdot\frac{\partial \Delta}{\partial x^\mu} - 2(\Box - m^2)\Delta^1\frac{\partial \Delta^1}{\partial x^\mu} = 0\,. \qquad \text{Q.E.D.}$$

A similar expression appears in the vacuum polarization in the presence of an external field (see Sec. 8): Define

$$\left.\begin{aligned} &\tfrac{1}{2}\varepsilon(x-x')\langle[j^\mu(x),\, j^\nu(x')]\rangle_0 = +ie^2\bar{K}_{\mu\nu}(x-x') \\ &\bar{K}_{\mu\nu}(x-x') = -\varepsilon(x-x')K_{\mu\nu}(x-x') \end{aligned}\right\}. \qquad [7.14]$$

Assertion: ε can be taken into the differentiation; i.e., $K_{\mu\nu}$ goes over into $\bar{K}_{\mu\nu}$ if Δ is replaced by $2\bar{\Delta}$, that is, according to [7.6],

$$\bar{K}_{\mu\nu}(\xi) = 4\left\{\frac{\partial \bar{\Delta}}{\partial \xi^\mu}\frac{\partial \Delta^1}{\partial \xi^\nu} + \frac{\partial \Delta^1}{\partial \xi^\mu}\frac{\partial \bar{\Delta}}{\partial \xi^\nu} - \right.$$

$$\left. - \delta_{\mu\nu}\left[\frac{\partial \bar{\Delta}}{\partial \xi^\varrho}\frac{\partial \Delta^1}{\partial \xi^\varrho} + m^2\bar{\Delta}\Delta^1\right]\right\}. \qquad [7.15]$$

Our assertion means that

$$\Delta\frac{\partial \varepsilon}{\partial x^\mu}\frac{\partial \Delta^1}{\partial x^\nu} + \frac{\partial \Delta^1}{\partial x^\mu}\frac{\partial \varepsilon}{\partial x^\nu}\Delta - \delta_{\mu\nu}\frac{\partial \varepsilon}{\partial x^\varrho}\frac{\partial \Delta^1}{\partial x^\varrho}\Delta = 0\,;$$

that is,

$$\delta_{\mu 4}\delta(t-t')\Delta\frac{\partial \Delta^1}{\partial x^\nu} + \delta_{\nu 4}\delta(t-t')\Delta\frac{\partial \Delta^1}{\partial x^\mu} - \delta_{\mu\nu}\delta(t-t')\frac{\partial \Delta^1}{\partial x^4}\Delta = 0\,.$$

Now, however, $\Delta(\boldsymbol{x}-\boldsymbol{x}', 0) = 0$, so that one can, with a certain amount of justification, assume this equivalence; it is only at the origin that it may not be valid. On the other hand, $\partial \bar{K}_{\mu\nu}/\partial x^\nu = 0$ is more critical. Since

$$(\Box - m^2)\bar{\Delta}(x) = -\delta^4(x) ,$$

then

$$\frac{\partial \bar{K}_{\mu\nu}}{\partial x^\nu} = 4\frac{\partial \Delta^1}{\partial x^\mu}(\Box - m^2)\bar{\Delta} = -4\delta^4(x)\frac{\partial \Delta^1(x)}{\partial x^\mu} .$$

Now, since

$$\Delta^1 = \frac{1}{2\pi^2(xx)} + \frac{m^2}{8\pi^2}\log|(xx)| + f_{\text{reg}}(xx)$$

is the expansion in the vicinity of the light cone, where $f_{\text{reg}}(-\lambda)$ is a regular function of λ, then

$$\frac{\partial \Delta^1}{\partial x^\mu} = -\frac{x^\mu}{\pi^2(xx)^2} + \frac{m^2}{4\pi^2}\frac{x^\mu}{(xx)} + 2x^\mu f'_{\text{reg}}(xx) .$$

This, therefore, does not vanish at the origin. Thus, $\partial \bar{K}_{\mu\nu}/\partial x^\nu$ is indeterminate.

In momentum space

$$\bar{\Delta} = \left(\frac{1}{2\pi}\right)^4 \mathscr{P}\int \frac{\exp[i(px)]}{p^2+m^2}\,\mathrm{d}^4p ,$$

so that

$$\bar{K}_{\mu\nu}(p) = 4\left(\frac{1}{2\pi}\right)^3\int \frac{\delta(k^2+m^2)}{(p-k)^2+m^2}[-k_\mu(p-k)_\nu - k_\nu(p-k)_\mu - \delta_{\mu\nu}(-(p-k)_\lambda k_\lambda + m^2)]\,\mathrm{d}^4k . \qquad [7.16]$$

Finally, one can also consider the kernel

$$\langle P(j^\mu(x), j^\nu(x'))\rangle_0 = e^2 K^C_{\mu\nu}(x-x') . \qquad [7.17]$$

Then, because

$$P(A(x), B(x')) = \tfrac{1}{2}\{A(x), B(x')\} + \tfrac{1}{2}\varepsilon(x-x')[A(x), B(x')] ,$$

one obtains

$$K^C_{\mu\nu} = \hat{K}_{\mu\nu} + i\bar{K}_{\mu\nu} . \qquad [7.18]$$

Furthermore, we also have

$$K^C_{\mu\nu}(x-x') = \frac{1}{e^2}\langle P(j^\mu(x), j^\nu(x'))\rangle_0$$

$$= \gamma^\mu_{\alpha\beta}\gamma^\nu_{\varrho\sigma}\langle P(\bar\psi_\beta(x), \psi_\varrho(x'))\,\varepsilon(x-x')\rangle_0$$

$$\cdot\langle P(\psi_\alpha(x), \bar\psi_\sigma(x'))\,\varepsilon(x-x')\rangle_0$$

$$= \tfrac{1}{4}\,\mathrm{Tr}\,\{\gamma^\mu S^C(x-x')\gamma^\nu S^C(x'-x)\};$$

hence

$$K^C_{\mu\nu} = -2\frac{\partial\Delta^C}{\partial x^\mu}\frac{\partial\Delta^C}{\partial x^\nu} + \delta_{\mu\nu}\left[+\frac{\partial\Delta^C}{\partial x^\varrho}\frac{\partial\Delta^C}{\partial x^\varrho} + m^2\Delta^C\Delta^C\right], \qquad [7.19]$$

which, because $\Delta^C = \Delta^1 - 2i\bar\Delta$, agrees with $\hat K_{\mu\nu} + i\bar K_{\mu\nu}$. (See the note below.)

We get

$$\frac{\partial K^C_{\mu\nu}}{\partial x^\nu} = -2\frac{\partial\Delta^C}{\partial x^\mu}(\Box - m^2)\Delta^C = -4i\frac{\partial\Delta^C}{\partial x^\mu}\delta^4(x)\,.$$

Here,

$$\Delta^C(x) = \lim_{\mu\to 0}\frac{-2i}{(2\pi)^4}\int\frac{\exp[i(kx)]}{k^2+m^2-i\mu^2}\,\mathrm{d}^4k\,,$$

and, in momentum space,

$$K^C_{\mu\nu}(p) = \frac{4}{(2\pi)^4}\int\frac{1}{k^2+m^2-i\mu^2}\cdot\frac{1}{(k-p)^2+m^2-i\mu^2}$$

$$\cdot[-k_\mu(p-k)_\nu - k_\nu(p-k)_\mu - \delta_{\mu\nu}(-(p_\lambda-k_\lambda)k_\lambda + m^2)]\,\mathrm{d}^4k\,.$$

Note: 1. Formally, all of this looks very nice; however, upon calculating one finds that

$$K^C_{\mu\nu} \neq \hat K_{\mu\nu} + i\bar K_{\mu\nu}\,.$$

For example, the real part of $K^C_{\mu\nu}$ is

$$\mathrm{Re}\,K^C_{\mu\nu} = 2\frac{\partial\bar\Delta}{\partial x^\mu}\frac{\partial\bar\Delta}{\partial x^\nu} - \delta_{\mu\nu}\left(\frac{\partial\bar\Delta}{\partial x^\varrho}\frac{\partial\bar\Delta}{\partial x^\varrho} + m^2\bar\Delta\bar\Delta\right) + \text{terms in } \Delta^1,$$

while

$$\hat K_{\mu\nu} = 2\frac{\partial\Delta}{\partial x^\nu}\frac{\partial\Delta}{\partial x^\mu} - \delta_{\mu\nu}\left(\frac{\partial\Delta}{\partial x^\varrho}\frac{\partial\Delta}{\partial x^\varrho} + m^2\Delta\Delta\right) + \text{terms in } \Delta^1,$$

and these two expressions differ in terms of the form

$$\left(\frac{\partial\varepsilon}{\partial x^\nu}\Delta\right)\left(\frac{\partial\varepsilon}{\partial x^\mu}\Delta\right)+\left(\frac{\partial\varepsilon}{\partial x^\nu}\frac{\partial\Delta}{\partial x^\mu}+\frac{\partial\varepsilon}{\partial x^\mu}\frac{\partial\Delta}{\partial x^\nu}\right)\varepsilon\Delta$$

$$-\delta_{\mu\nu}\cdot(\text{corresponding terms})\,.$$

Now, we have indeed postulated that $(\partial\varepsilon/\partial x^\nu)\Delta = 0$, so that from $\bar{S} = -\frac{1}{2}\varepsilon S$ and $S = (\gamma(\partial/\partial x) - m)\Delta$ it follows that

$$\bar{S} = \left(\gamma\frac{\partial}{\partial x} - m\right)\bar{\Delta}\,.$$

However, the expressions $(\partial\varepsilon/\partial x^\mu)(\partial\Delta/\partial x^\nu)\varepsilon\Delta$ are obviously of the type

$$\delta_{\mu 4}\delta_{\nu 4}\delta^4(x)\,\bar{\Delta} = \bar{\Delta}(0)\,,$$

which is singular. They are, therefore, indeterminate. This difficulty arises from the fact that initially it is always postulated that $(\partial\varepsilon/\partial x^\nu)\Delta = 0$, but this may not be used here since this quantity is multiplied by a singular function.

Such difficulties can be removed with the regularization procedure of Pauli and Villars,[1] a formal procedure for avoiding divergences. One sets

$$(K_{\mu\nu}(x))_{\text{reg}} = \sum_{i=0}^{N} C_i K_{\mu\nu}(x_i M_i^2)\,,$$

where

$$C_0 = 1\,,\qquad M_0 = m\,,$$

$$\sum_{i=0}^{N} C_i = 0\,,\qquad \sum_{i=0}^{N} C_i M_i^2 = 0\,,$$

and, at the end, one lets $M_i \to \infty$ $(i \neq 0)$. The physical expressions then become finite, and difficulties of the kind mentioned above no longer occur.

2. The $K^C_{\mu\nu}$ is, however, somewhat peculiar since $\hat{K}_{\mu\nu}$ is entirely regular and is replaced by a singular quantity when $K^C_{\mu\nu}$ is used.

[1] W. Pauli and F. Villars, *Rev. Mod. Phys.* **21**, 434 (1949).

8. VACUUM POLARIZATION IN AN EXTERNAL FIELD [2]

In the presence of an external field $\mathscr{A}_\nu$, the Dirac equations read

$$\left(\gamma^\nu \frac{\partial}{\partial x^\nu} + m\right)\psi(x) = ie\gamma^\nu \mathscr{A}_\nu(x)\psi(x);$$

$$\bar{\psi}(x)\left(\gamma^\nu \frac{\partial}{\partial x^\nu} - m\right) = -ie\bar{\psi}(x)\gamma^\nu \mathscr{A}_\nu(x).$$

If one introduces the expansion

$$\psi(x) = \psi^{(0)}(x) + e\psi^{(1)}(x) + \dots,$$

where $\psi^{(n)}(x) \to 0$ for $t \to -\infty$ $(n \neq 0)$, and where $\psi^{(0)}$ is a solution of the field-free Dirac equation, then for $\psi^{(1)}$ there follows the differential equation

$$\left(\gamma^\nu \frac{\partial}{\partial x^\nu} + m\right) e\psi^{(1)} = ie\gamma^\nu \mathscr{A}_\nu \psi^{(0)}.$$

With the help of the Green's functions

$$S^{\text{ret}} = \bar{S} - \tfrac{1}{2}S, \qquad S^{\text{adv}} = \bar{S} + \tfrac{1}{2}S,$$

the solutions are

$$\left.\begin{aligned} e\psi^{(1)} &= -ie\int S^{\text{ret}}(x-x')\gamma^\nu \mathscr{A}_\nu(x')\psi^{(0)}(x')\,\mathrm{d}^4x' \\ e\bar{\psi}^{(1)} &= -ie\int \bar{\psi}^{(0)}(x')\gamma^\nu \mathscr{A}_\nu(x')S^{\text{adv}}(x'-x)\,\mathrm{d}^4x' \end{aligned}\right\}. \qquad [8.1]$$

This can be easily verified by substituting into the differential equation. We get, for example,

$$\begin{aligned} \left(\gamma^\nu \frac{\partial}{\partial x^\nu} + m\right) S^{\text{ret}} &= \left(\gamma \frac{\partial}{\partial x} + m\right)\left(\gamma \frac{\partial}{\partial x} - m\right)(\bar{\Delta} - \tfrac{1}{2}\Delta) \\ &= (\Box - m^2)(\bar{\Delta} - \tfrac{1}{2}\Delta) = -\delta^4(x). \end{aligned}$$

[2] G. Källén, *Helv. Phys. Acta* **22**, 637 (1949).

If the expansion is substituted into

$$j^{\mu}(x) = -\frac{ie}{2}\,[\psi_\alpha(x), \bar{\psi}_\beta(x)]\gamma^{\mu}_{\alpha\beta}\,,$$

then

$$j^{\mu(1)}(x) = -\frac{ie^2}{2}\left([\psi^{(1)}_\alpha(x), \bar{\psi}^{(0)}_\beta(x)] + [\psi^{(0)}_\alpha(x), \bar{\psi}^{(1)}_\beta(x)]\right)\gamma^{\mu}_{\alpha\beta}\,,$$

and, with Eq. [8.1],

$$\langle j^{\mu(1)}(x)\rangle_0 = \frac{e^2}{2}\int \mathscr{A}_\nu(x')\,\mathrm{Tr}\,\{\gamma^\mu S^{\mathrm{ret}}(x-x')\gamma^\nu S^1(x'-x) + \gamma^\mu S^1(x-x')\gamma^\nu S^{\mathrm{adv}}(x'-x)\}\,\mathrm{d}^4x'\,. \quad [8.2]$$

Here, if S^{ret} and S^{adv} are replaced by the expressions with $\bar{S}$ and S, upon explicitly calculating the trace one obtains

$$\langle j^{\mu(1)}(x)\rangle_0 = e^2\int [K_{\mu\nu}(x-x') - \bar{K}_{\mu\nu}(x-x')]\,\mathscr{A}_\nu(x')\,\mathrm{d}^4x'\,,$$

$$\langle j^{\mu(1)}(x)\rangle_0 = i\int \langle [j^\mu(x), j^\nu(x')]\rangle_0\,\frac{1+\varepsilon(x-x')}{2}\,\mathscr{A}_\nu(x')\,\mathrm{d}^4x'\,. \quad [8.3]$$

This method does not employ the concept of energy. It will prove to be convenient in the treatment of spin-zero particles.

9. PARTICLES WITH ZERO SPIN

These particles are described by means of a complex scalar field Φ which satisfies the differential equation

$$(\Box - m^2)\,\Phi(x) = 0\,. \quad [9.1]$$

In the c-number theory, the current is written as

$$j^\nu = ie\left(\frac{\partial \Phi^*}{\partial x^\nu}\,\Phi - \frac{\partial \Phi}{\partial x^\nu}\,\Phi^*\right), \quad [9.2]$$

for which $\partial j^\nu/\partial x^\nu = 0$ follows.

Let Φ be expanded in terms of a complete set of eigenfunctions:

$$\Phi(x) = \sum_r A_r u^{(r)}(x) , \qquad [9.3]$$

where, for example, in the case of plane waves,

$$u^{(r)}(x) = \frac{1}{\sqrt{G}} \frac{1}{\sqrt{2\omega_r}} \exp[i(k_r x)] , \quad \omega_r = +\sqrt{k_r^2 + m^2} ,$$
$$k_0 = \pm \omega .$$

Then, the charge is

$$e = \int j^0 \mathrm{d}^3x ; \quad e_{rs} = -ie \int \left(\frac{\partial u^{(r)*}}{\partial x^0} u^{(s)} - \frac{\partial u^{(s)}}{\partial x^0} u^{(r)*} \right) \mathrm{d}^3x = e\varepsilon_r \delta_{rs}$$

and the energy is

$$E_{rs} = \int \left(\frac{\partial u^{(r)*}}{\partial x^0} \frac{\partial u^{(s)}}{\partial x^0} + \frac{\partial u^{(r)*}}{\partial \boldsymbol{x}} \cdot \frac{\partial u^{(s)}}{\partial \boldsymbol{x}} + m^2 u^{(r)*} u^{(s)} \right) \mathrm{d}^3x = \omega_r \delta_{rs} .$$

(These last expressions are valid if we make E_{rs} diagonal.) One recognizes that here the energy is positive definite while the charge is not. The following completeness relations hold:

$$\left. \begin{aligned} \sum_r u^{(r)}(x) u^{(r)*}(x') \varepsilon_r &= i\Delta(x - x') \\ \sum_r u^{(r)}(x) u^{(r)*}(x') &= \Delta^1(x - x') \end{aligned} \right\} . \qquad [9.4]$$

For quantization, we must invoke Bose-Einstein statistics, i.e., we use the commutators (instead of the anticommutators, which were used for spin-$\frac{1}{2}$ particles). We see that we can proceed without difficulties only if we set

$$[A_r, A_s^*] = \varepsilon_r \cdot \delta_{rs} ; \qquad [9.5]$$

i.e.,

$$A_r^* A_r = N_r \qquad (\varepsilon_r > 0) ,$$
$$A_r A_r^* = N_r \qquad (\varepsilon_r < 0) ,$$

or

$$A_r^* A_r = N_r \frac{1+\varepsilon_r}{2} + (N_r + 1) \frac{1-\varepsilon_r}{2} ,$$

$$\left. \begin{aligned} A_r^* A_r &= N_r + \frac{1-\varepsilon_r}{2} \\ A_r A_r^* &= N_r + \frac{1+\varepsilon_r}{2} \end{aligned} \right\} . \qquad [9.6]$$

From this it follows, then, that

$$[\Phi(x), \Phi^*(x')] = i\Delta(x - x') , \qquad [9.7]$$

$$\langle\{\Phi(x), \Phi^*(x')\}\rangle_0 = \Delta^1(x - x') . \qquad [9.8]$$

Note that here Φ and Φ^* can be freely interchanged:

$$\left.\begin{aligned} [\Phi^*(x), \Phi(x')] &= [\Phi(x), \Phi^*(x')] = i\Delta(x - x') \\ \langle\{\Phi^*(x), \Phi(x')\}\rangle_0 &= \langle\{\Phi(x), \Phi^*(x')\}\rangle_0 = \Delta^1(x - x') \end{aligned}\right\} . \qquad [9.9]$$

Consequently, charge conjugation becomes very trivial here:

$$\left.\begin{aligned} \Phi' &= \Phi^* \\ \Phi^{*\prime} &= \Phi \end{aligned}\right\} . \qquad [9.10]$$

Thus, even in the c-number theory, the current changes sign under this transformation. In the q-number theory, one must first symmetrize appropriately:

$$j^\nu(x) = \frac{ie}{2}\left(\left\{\frac{\partial\Phi^*}{\partial x^\nu}, \Phi\right\} - \left\{\frac{\partial\Phi}{\partial x^\nu}, \Phi^*\right\}\right), \qquad [9.11]$$

where $\{A(x), B(x)\}$ is understood to be

$$\lim_{x' \to x} \tfrac{1}{2}\left(\{A(x'), B(x)\} + \{A(x), B(x')\}\right) .$$

Then, $j'^\nu(x) = -j^\nu(x)$ even before the limiting process, and thus

$$\langle j^\nu(x)\rangle_0 = 0 .$$

Note: In analogy to the c-number theory, the energy would be

$$E = \int\left(\frac{\partial\Phi^*}{\partial x^0}\frac{\partial\Phi}{\partial x^0} + \frac{\partial\Phi^*}{\partial \boldsymbol{x}}\cdot\frac{\partial\Phi}{\partial \boldsymbol{x}} + m^2\Phi^*\Phi\right) d^3x \qquad [9.12]$$

$$= \sum_r \omega_r A_r^* A_r = \sum_r \omega_r\left(N_r + \frac{1 - \varepsilon_r}{2}\right)$$

in so far as we may set $\sum_r \varepsilon_r = 0$; or, had we taken the anti-commutator above, this would become

$$E = \sum_r (N_r + \tfrac{1}{2})\omega_r = \sum_r \omega_r(N_r^+ + N_r^- + 1) .$$

For electrons we had

$$E = \sum_{r.\lambda} \omega_r(N_r^+ + N_r^- - 1) , \qquad \lambda = 1, 2 .$$

One can ask whether these zero-point energies can compensate one another. We have

$$\begin{cases} \text{spin } 0: & \dfrac{E_0}{V} = \left(\dfrac{1}{2\pi}\right)^3 \displaystyle\int \sqrt{k^2+m^2}\, \mathrm{d}^3k; \\ \text{spin } \frac{1}{2}: & \dfrac{E_0}{V} = -2\left(\dfrac{1}{2\pi}\right)^3 \displaystyle\int \sqrt{k^2+m^2}\, \mathrm{d}^3k; \end{cases}$$

where the masses are, in general, different. For compensation we must calculate

$$\int_0^K k^2 \sqrt{k^2+m^2}\, \mathrm{d}k = \frac{K^4}{4} + \frac{1}{4} m^2K^2 - \frac{m^4}{4} \log \frac{2K}{m} + O\left(\frac{1}{K}\right).$$

We see that the compensation requirements are

$$\begin{pmatrix}\text{the number of kinds of} \\ \text{spin-zero particles}\end{pmatrix} = 2 \times \begin{pmatrix}\text{the number of kinds} \\ \text{of spin-}\frac{1}{2}\text{ particles}\end{pmatrix};$$

i.e.,

$$Z_0 = 2\, Z_{\frac{1}{2}}\, .$$

Furthermore,

$$\sum_i (m_0^i)^2 = 2 \sum (m_{\frac{1}{2}}^i)^2\, ,$$

$$\sum_i (m_0^i)^4 = 2 \sum (m_{\frac{1}{2}}^i)^4\, ,$$

$$\sum (m_0^i)^4 \log m_0^i = 2 \sum_i (m_{\frac{1}{2}}^i)^4 \log m_{\frac{1}{2}}^i\, .$$

These requirements are so extensive that it is rather improbable that they are satisfied in reality.

Some additional formulas are

$$\left.\begin{aligned} \langle \Phi(x)\, \Phi^*(x')\rangle_0 &= \langle \Phi^*(x)\, \Phi(x')\rangle_0 \\ &= \tfrac{1}{2}(\Delta^1 + i\Delta)(x-x') = +i\Delta^+(x-x') \\ \langle \Phi^*(x')\, \Phi(x)\rangle_0 &= \langle \Phi(x')\, \Phi^*(x)\rangle_0 \\ &= \tfrac{1}{2}(\Delta^1 - i\Delta)(x-x') = -i\Delta^-(x-x') \\ \langle P(\Phi^*(x)\, \Phi(x'))\rangle_0 &= \tfrac{1}{2}\Delta^C(x-x') \end{aligned}\right\}. \qquad [9.13]$$

Remark concerning the current: In an external field $\mathscr{A}_\mu$, an additional term must be added to Eq. [9.11] in accordance with the general rule $\partial/\partial x^\mu \to \partial/\partial x^\mu + ie\mathscr{A}_\mu$:

$$j^\mu(x) = \frac{ie}{2}\left(\left\{\frac{\partial\Phi^*}{\partial x^\mu}, \Phi(x)\right\} - \left\{\Phi^*(x), \frac{\partial\Phi}{\partial x^\mu}\right\}\right) - \frac{1}{2}e^2\mathscr{A}_\mu(x)\{\Phi^*(x), \Phi(x)\}. \qquad [9.14]$$

The continuity equation then remains valid.

In forming the vacuum expectation values of expressions bilinear in the current, one has to form the vacuum expectation values of expressions such as

$$\Phi(x)\frac{\partial\Phi^*(x)}{\partial x^\mu}\Phi(x')\frac{\partial\Phi^*(x')}{\partial x'^\nu}.$$

Here, one has again only to take pairs of fields with different variables x and x' and thus can omit the anticommutators (compare the spin-$\frac{1}{2}$ case):

$$\begin{aligned}\langle j^\mu(x)j^\nu(x')\rangle_0 = &-e^2\left\langle\frac{\partial\Phi^*(x)}{\partial x^\mu}\Phi(x')\right\rangle_0\left\langle\Phi(x)\frac{\partial\Phi^*(x')}{\partial x'^\nu}\right\rangle_0 \\ &+e^2\left\langle\frac{\partial\Phi^*(x)}{\partial x^\mu}\frac{\partial\Phi(x')}{\partial x'^\nu}\right\rangle_0\langle\Phi(x)\,\Phi^*(x')\rangle_0 \\ &-e^2\left\langle\frac{\partial\Phi(x)}{\partial x^\mu}\Phi^*(x')\right\rangle_0\left\langle\Phi^*(x)\frac{\partial\Phi(x')}{\partial x'^\nu}\right\rangle_0 \\ &+e^2\left\langle\frac{\partial\Phi(x)}{\partial x^\mu}\frac{\partial\Phi^*(x')}{\partial x'^\nu}\right\rangle_0\langle\Phi^*(x)\,\Phi(x')\rangle_0 + O(e^3),\end{aligned}$$

$$\langle j^\mu(x)j^\nu(x')\rangle_0 = 2e^2\left(-\frac{\partial\Delta^+}{\partial x^\mu}\frac{\partial\Delta^+}{\partial x^\nu} + \frac{\partial^2\Delta^+}{\partial x^\mu\partial x^\nu}\Delta^+\right) + O(e^3). \qquad [9.15]$$

If one drops the $O(e^3)$ terms (which is equivalent to setting $\mathscr{A}_\mu = 0$) and defines

$$\tfrac{1}{2}\langle\{j^\mu(x), j^\nu(x')\}\rangle_0 \equiv e^2\hat{L}_{\mu\nu}(x-x'), \qquad [9.16]$$

$$\tfrac{1}{2}\langle[j^\mu(x), j^\nu(x')]\rangle_0 \equiv -ie^2L_{\mu\nu}(x-x'), \qquad [9.17]$$

then

$$\hat{L}_{\mu\nu}(\xi) = \frac{1}{2}\left(\frac{\partial\Delta^1}{\partial\xi^\mu}\frac{\partial\Delta^1}{\partial\xi^\nu} - \frac{\partial\Delta}{\partial\xi^\mu}\frac{\partial\Delta}{\partial\xi^\nu} - \frac{\partial^2\Delta^1}{\partial\xi^\mu\partial\xi^\nu}\Delta^1 + \frac{\partial^2\Delta}{\partial\xi^\mu\partial\xi^\nu}\Delta\right), \qquad [9.18]$$

$$L_{\mu\nu}(\xi) = -\frac{1}{2}\left(\frac{\partial\Delta}{\partial\xi^\mu}\frac{\partial\Delta^1}{\partial\xi^\nu} + \frac{\partial\Delta^1}{\partial\xi^\mu}\frac{\partial\Delta}{\partial\xi^\nu} - \Delta\frac{\partial^2\Delta^1}{\partial\xi^\mu\partial\xi^\nu} - \Delta^1\frac{\partial^2\Delta}{\partial\xi^\mu\partial\xi^\nu}\right). \qquad [9.19]$$

In analogy to Section 8, we can also calculate the vacuum polarization in an external field:

$$\Phi = \Phi^{(0)} + \Phi^{(1)} + \dots ,$$

$$\Phi^{(1)} = 0 \qquad \text{for } t \to -\infty ,$$

$$(\Box - m^2)\,\Phi^{(0)} = 0 ,$$

$$\Phi^{(1)}(x) = -ie\int \mathscr{A}_\nu(x')\left(\Delta^{\text{ret}}(x-x')\,\frac{\partial\Phi^{(0)}(x')}{\partial x'^\nu} + \frac{\partial\Delta^{\text{ret}}(x-x')}{\partial x^\nu}\,\Phi^{(0)}(x')\right)\mathrm{d}^4x' . \qquad [9.20]$$

Thus,

$$j^{\mu(1)}(x) = \frac{ie}{2}\left(\left\{\frac{\partial\Phi^{(\nu)*}}{\partial x^\mu}, \Phi^{(1)}\right\} - \left\{\Phi^{(\nu)*}, \frac{\partial\Phi^{(1)}}{\partial x^\mu}\right\} + \left\{\frac{\partial\Phi^{(1)*}}{\partial x^\mu}, \Phi^{(0)}\right\} - \left\{\Phi^{(1)*}, \frac{\partial\Phi^{(0)}}{\partial x^\mu}\right\}\right) - \frac{1}{2}\,e^2\mathscr{A}_\mu\{\Phi^{(0)*}, \Phi^{(0)}\} ,$$

$$\begin{aligned}\langle j^{\mu(1)}(x)\rangle_0 = \frac{e^2}{2}\int \mathscr{A}_\nu(x')\Bigg[&\Delta^{\text{ret}}(x-x')\left\langle\left\{\frac{\partial\Phi^{(0)*}(x)}{\partial x^\mu}, \frac{\partial\Phi^{(0)}(x')}{\partial x'^\nu}\right\}\right\rangle_0 \\ &+ \frac{\partial\Delta^{\text{ret}}(x-x')}{\partial x^\nu}\left\langle\left\{\frac{\partial\Phi^{(0)*}(x)}{\partial x^\mu}, \Phi^{(0)}(x')\right\}\right\rangle_0 \\ &- \frac{\partial\Delta^{\text{ret}}(x-x')}{\partial x^\mu}\left\langle\left\{\Phi^{(0)*}(x), \frac{\partial\Phi^{(0)}(x')}{\partial x'^\nu}\right\}\right\rangle_0 \\ &- \frac{\partial^2\Delta^{\text{ret}}(x-x')}{\partial x^\mu\partial x^\nu}\langle\{\Phi^{(0)*}(x), \Phi^{(0)}(x')\}\rangle_0 \\ &- \frac{\partial\Delta^{\text{ret}}(x-x')}{\partial x^\mu}\left\langle\left\{\frac{\partial\Phi^{(0)*}(x')}{\partial x'^\nu}, \Phi^{(0)}(x)\right\}\right\rangle_0 \\ &- \frac{\partial^2\Delta^{\text{ret}}(x-x')}{\partial x^\mu\partial x^\nu}\langle\{\Phi^{(0)*}(x), \Phi^{(0)}(x')\}\rangle_0 \\ &+ \Delta^{\text{ret}}(x-x')\left\langle\left\{\frac{\partial\Phi^{(0)*}(x')}{\partial x'^\nu}, \frac{\partial\Phi^{(0)}(x)}{\partial x^\mu}\right\}\right\rangle_0 \\ &+ \frac{\partial\Delta^{\text{ret}}(x-x')}{\partial x^\nu}\left\langle\left\{\Phi^{(0)*}(x'), \frac{\partial\Phi^{(0)}(x)}{\partial x^\mu}\right\}\right\rangle_0 \\ &- \delta_{\mu\nu}\delta^4(x-x')\langle\{\Phi^{(0)*}(x), \Phi^{(0)}(x')\}\rangle_0\Bigg]\,\mathrm{d}^4x' .\end{aligned}$$

If one substitutes $\Delta^{\text{ret}} = \bar{\Delta} - \frac{1}{2}\Delta$, then it is seen that one can write

$$\langle j^{\mu(1)}(x)\rangle_0 = \frac{i}{2}\int \mathscr{A}_\nu(x')\langle[j^{\mu(0)}(x), j^{\nu(0)}(x')]\rangle_0\, \mathrm{d}^4x' - e^2\int \mathscr{A}_\nu(x')\bar{L}_{\mu\nu}(x-x')\,\mathrm{d}^4x', \qquad [9.21]$$

where

$$\bar{L}_{\mu\nu}(\xi) = -\left(\frac{\partial\bar{\Delta}}{\partial\xi^\mu}\frac{\partial\Delta^1}{\partial\xi^\nu} + \frac{\partial\bar{\Delta}}{\partial\xi^\nu}\frac{\partial\Delta^1}{\partial\xi^\mu} - \bar{\Delta}\frac{\partial^2\Delta^1}{\partial\xi^\mu\partial\xi^\nu} - \Delta^1\frac{\partial^2\bar{\Delta}}{\partial\xi^\mu\partial\xi^\nu} - \delta_{\mu\nu}\delta^4(\xi)\Delta^1\right). \qquad [9.22]$$

Note: 1.
$$\bar{L}_{\mu\nu}(\xi) \neq \varepsilon(\xi)\cdot L_{\mu\nu}(\xi).$$

2. One can define

$$L^C_{\mu\nu} = \hat{L}_{\mu\nu} + i\bar{L}_{\mu\nu}.$$

However, it is again *not* true that

$$\langle P(j^\mu(x)\, j^\nu(x'))\rangle_0 = L^C_{\mu\nu}(x-x'),$$

since $\bar{L}_{\mu\nu} \neq \varepsilon L_{\mu\nu}$. On the contrary,

$$L^C_{\mu\nu}(\xi) = \frac{1}{2}\left(\frac{\partial\Delta^C}{\partial\xi^\mu}\frac{\partial\Delta^C}{\partial\xi^\nu} - \frac{\partial^2\Delta^C}{\partial\xi^\mu\partial\xi^\nu}\Delta^C + 2i\delta^4(\xi)\Delta^1(\xi)\delta_{\mu\nu}\right). \qquad [9.23]$$

3. The quantity $\langle j^{\mu(1)}\rangle_0$ should satisfy the continuity equation in order that gauge invariance be satisfied. A calculation shows that this is not so.

In Section 7 we had

$$\frac{\partial\bar{K}_{\mu\nu}}{\partial x^\nu} = -4\delta^4(x)\frac{\partial\Delta^1(x)}{\partial x^\mu}. \qquad [9.24]$$

Correspondingly, one finds

$$\frac{\partial\bar{L}_{\mu\nu}}{\partial x^\nu} = +2\delta^4(x)\frac{\partial\Delta^1(x)}{\partial x^\mu}. \qquad [9.25]$$

4. Mixtures of spin-0 and spin-$\frac{1}{2}$ particles (Rayski, Umezawa [3]). For the case of equal particle masses, the presence of twice as many spin-0 particles as spin-$\frac{1}{2}$ particles suffices for the com-

[3] J. Rayski, *Acta phys. Polonica* **9**, 129 (1948); H. Umezawa, J. Yukawa, and E. Yamada, *Progr. Theor. Phys.* **3**, 317 (1948).

pensation of the indeterminate expressions of [9.24] and [9.25]. For different masses it is sufficient that

$$N_0 = 2N_{\frac{1}{2}}$$
$$\sum_i (m_0^i)^2 = 2\sum_i (m_{\frac{1}{2}}^i)^2 .$$

These conditions are contained in the conditions for the compensation of the zero-point energy (see p. 33). For different masses, since

$$\frac{\partial \Delta^1}{\partial x^\mu} \sim \frac{x^\mu}{(xx)^2} + am^2 \frac{x^\mu}{(xx)} + x^\mu f_{\text{reg}}(xx) ,$$

the compensation of the second term also yields a condition. In any event, it is questionable whether this mixture has physical significance.

5. The introduction of the Δ^C functions by the groupings $K^C = \hat{K} + i\bar{K}$ and $L^C = \hat{L} + i\bar{L}$ is artificial here; it first becomes meaningful in the S matrix. As with $K^C_{\mu\nu}$, artificial singularities of the form $\delta^4(x)\bar{\Delta}(x)$ are also introduced with $L^C_{\mu\nu}$.

With

$$L(p) = \int \exp[-ipx] L(x)\, \mathrm{d}^4x ,$$

it is immediately found that in momentum space

$$L^C_{\mu\nu}(p) = -\left(\frac{1}{2\pi}\right)^4 \int [-k_\mu(p-k)_\nu - k_\nu(p-k)_\mu$$
$$+ (p-k)_\mu(p-k)_\nu + k_\mu k_\nu - \delta_{\mu\nu}(k^2 + 2m^2 + (p-k)^2)]$$
$$\cdot \frac{\mathrm{d}^4k}{(k^2+m^2-i\mu^2)((k-p)^2+m^2-i\mu^2)} , \qquad [9.26]$$

$$\hat{L}_{\mu\nu}(p) = \frac{1}{4}\left(\frac{1}{2\pi}\right)^2 \int [-k_\mu(p-k)_\nu - k_\nu(p-k)_\mu$$
$$+ (p-k)_\mu(p-k)_\nu + k_\mu k_\nu][1 + \varepsilon(k)\varepsilon(p-k)]$$
$$\cdot \delta(k^2+m^2)\delta((k-p)^2+m^2)\, \mathrm{d}^4k , \qquad [9.27]$$

$$\bar{L}_{\mu\nu}(p) = \frac{i}{4}\left(\frac{1}{2\pi}\right)^2 \int [-k_\mu(p-k)_\nu - k_\nu(p-k)_\mu$$
$$+ (p-k)_\mu(p-k)_\nu + k_\mu k_\nu][\varepsilon(k) + \varepsilon(p-k)]$$
$$\cdot \delta(k^2+m^2)\delta((k-p)^2+m^2)\, \mathrm{d}^4k . \qquad [9.28]$$

Since

$$\varepsilon(k) + \varepsilon(p-k) = \varepsilon(p)[1 + \varepsilon(k)\varepsilon(p-k)],$$

we have

$$L_{\mu\nu}(p) = i\varepsilon(p)\hat{L}_{\mu\nu}(p).$$

10. EVALUATION OF THE KERNELS $\hat{K}$ AND $\hat{L}$

The quantities $\hat{K}_{\mu\nu}$ and $\hat{L}_{\mu\nu}$ are related to the charge fluctuations in the vacuum. We will now work these out. We have

$$\hat{K}_{\mu\nu}(p) = \left(\frac{1}{2\pi}\right)^2 \int \delta(k^2+m^2)\delta\big((k-p)^2+m^2\big)[\varepsilon(k)\varepsilon(p-k)+1]$$
$$\cdot[-2k_\mu k_\nu + k_\mu p_\nu + k_\nu p_\mu - \delta_{\mu\nu}\big((pk) - k^2 - m^2\big)]\,\mathrm{d}^4k,$$
$$\hat{L}_{\mu\nu}(p) = \left(\frac{1}{2\pi}\right)^2 \int \delta(k^2+m^2)\delta\big((k-p)^2+m^2\big)[\varepsilon(k)\varepsilon(p-k)+1]$$
$$\cdot\tfrac{1}{2}[2k_\mu k_\nu - k_\mu p_\nu - k_\nu p_\mu + \tfrac{1}{2}p_\mu p_\nu]\,\mathrm{d}^4k.$$

For the calculation, we note that the δ-functions simultaneously require that

$$k^2 + m^2 = 0$$

and that

$$p^2 - 2(pk) = 0.$$

This is possible for space-like as well as time-like p's.

1. *Space-like* p. In coordinates where $p = (\boldsymbol{p}, 0)$, we have $\boldsymbol{p}^2 - 2\boldsymbol{k}\cdot\boldsymbol{p} = 0$. Then, however,

$$\varepsilon(k)\varepsilon(p-k) = \varepsilon(-k)\varepsilon(k) = -1,$$

and the integrals vanish identically.

2. *Time-like* p. We choose $p = (\boldsymbol{0}, ip_0)$. Then, $-p_0^2 + 2p_0k_0 = 0$, so that $k_0 = \frac{1}{2}p_0 = \pm\sqrt{\boldsymbol{k}+{}^2m^2}$, and

$$\varepsilon(k)\varepsilon(p-k) = \varepsilon(k_0)\varepsilon(k_0) = +1.$$

Furthermore, $\boldsymbol{k}^2 = \frac{1}{4}p_0^2 - m^2$ so that $p_0^2 \geqslant 4m^2$, or, in general, $-p^2 \geqslant 4m^2$.

Now,

$$\hat{K}_{11}(p) = \left(\frac{1}{2\pi}\right)^2 \cdot 2 \int \delta(k^2 + m^2)\,\delta\big(p^2 - 2(kp)\big)\,[-2k_1^2 + p_0 k_0]\,\mathrm{d}^4k$$

$$= \left(\frac{1}{2\pi}\right)^2 \cdot 2 \int \delta(k^2 + m^2)\,\delta\big(p^2 - 2(kp)\big)\,[-\tfrac{2}{3}\boldsymbol{k}^2 + \tfrac{1}{2}p_0^2]\,\mathrm{d}^4k\,,$$

since

$$\begin{cases} k_1^2 \simeq \tfrac{1}{3}\boldsymbol{k}^2\,, \\ p_0 k_0 = +\tfrac{1}{2}p_0^2 \end{cases}$$

under the integral sign. But

$$-\frac{2}{3}\boldsymbol{k}^2 + \frac{1}{2}p_0^2 = -\frac{1}{6}p_0^2 + \frac{2}{3}m^2 + \frac{1}{2}p_0^2 = \frac{1}{3}(p_0^2 + 2m^2)\,,$$

so that

$$\hat{K}_{11}(p) = \left(\frac{1}{2\pi}\right)^2 \int \delta(k^2 + m^2)\,\delta\big(p^2 - 2(kp)\big) \cdot \tfrac{2}{3}(p_0^2 + 2m^2)\,\mathrm{d}^4k\,.$$

Furthermore

$$\begin{cases} \hat{K}_{44}(p) = 0 \\ \hat{L}_{44}(p) = 0 \end{cases}$$

(because $\hat{K}_{\mu\nu}\,p_\nu = 0$, $\hat{L}_{\mu\nu}\,p_\nu = 0$).

For an arbitrary coordinate system, we have yet to insert $(p_\mu p_\nu - \delta_{\mu\nu}p^2)/(-p^2)$, since, in our coordinate system, this becomes 1 for $\mu = \nu = 1$ and 0 for $\mu = \nu = 4$. Thus,

$$\left.\begin{matrix} \hat{K}_{\mu\nu}(p) \\ \hat{L}_{\mu\nu}(p) \end{matrix}\right\} = \left(\frac{1}{2\pi}\right)^2 \int \delta(k^2 + m^2)\,\delta\big((k-p)^2 + m^2\big)$$

$$\cdot 2 \cdot \frac{p_\mu p_\nu - \delta_{\mu\nu}p^2}{-p^2} \cdot \left\{\begin{matrix} \tfrac{1}{3}(-p^2 + 2m^2) \\ \tfrac{1}{3}(-\tfrac{1}{4}p^2 - m^2) \end{matrix}\right\} \mathrm{d}^4k\,.$$

The sign of k_0 is fixed ($k_0 = p_0/2$).

By using the fact that

$$\delta(f(z)) = \sum_\nu \frac{\delta(z - z_\nu)}{|f'(z_\nu)|},$$

where $f(z_\nu) = 0$, we get the following:

$$\int \delta(k^2 + m^2)\,\delta\big((k-p)^2 + m^2\big)\,\mathrm{d}^4k$$

$$= \int \delta\big(2\overset{+}{\sqrt{\boldsymbol{k}^2 + m^2}}|p_0| - 2\boldsymbol{k}\cdot\boldsymbol{p} - p_0^2 + \boldsymbol{p}^2\big)\,\frac{\mathrm{d}^3k}{2\overset{+}{\sqrt{\boldsymbol{k}^2 + m^2}}},$$

which for $\boldsymbol{p} = 0$ is equal to

$$4\pi \int \delta\big(2\overset{+}{\sqrt{k^2 + m^2}}|p_0| - p_0^2\big)\,\frac{k^2\mathrm{d}k}{2\overset{+}{\sqrt{k^2 + m^2}}}$$

$$= 2\pi\,\frac{k^2}{\overset{+}{\sqrt{k^2 + m^2}}}\cdot\frac{1}{\big(2k/(\overset{+}{\sqrt{k^2 + m^2}})\big)|p_0|}\Bigg|_{k^2 = \frac{1}{4}p_0^2 - m^2}$$

$$= \frac{\pi}{2}\cdot\sqrt{\frac{4m^2 + p^2}{p^2}}\,.$$

Thus,

$$\left.\begin{matrix}\hat{K}_{\mu\nu}(p)\\ \hat{L}_{\mu\nu}(p)\end{matrix}\right\} = \frac{1}{12\pi}\sqrt{\frac{4m^2 + p^2}{p^2}}$$

$$\cdot\frac{p_\mu p_\nu - \delta_{\mu\nu}p^2}{-p^2}\left\{\begin{matrix}(-p^2 + 2m^2)\\ (-\frac{1}{4}p^2 - m^2)\end{matrix}\right. \quad \text{for } -p^2 \geqslant 4m^2, \qquad [10.1]$$

$$\left.\begin{matrix}\hat{K}_{\mu\nu}(p)\\ \hat{L}_{\mu\nu}(p)\end{matrix}\right\} = 0 \qquad \text{otherwise}\,.$$

Now, we had

$$\left.\begin{matrix}\hat{K}_{\mu\nu}(x)\\ \hat{L}_{\mu\nu}(x)\end{matrix}\right\} = \left(\frac{1}{2\pi}\right)^4 \int \exp[i(px)]\left\{\begin{matrix}\hat{K}_{\mu\nu}(p)\\ \hat{L}_{\mu\nu}(p)\end{matrix}\right\}\mathrm{d}^4p$$

and

$$\left.\begin{matrix}\hat{K}_{\mu\nu}(x)\\ \hat{L}_{\mu\nu}(x)\end{matrix}\right\} = \frac{1}{2e^2}\langle\{j^\mu(x), j^\nu(x')\}\rangle_0\,,$$

where the upper part is valid for spin-$\frac{1}{2}$, the lower for spin-0. Equation [10.1] determines the charge fluctuations in a space-time region V:

$$\langle Q_V^2\rangle_0 = \left\langle\left(\int_V j^0(x)\,\mathrm{d}^4x\right)^2\right\rangle_0 = e^2\int \mathrm{d}^4x\int \mathrm{d}^4x'$$

$$\cdot\begin{Bmatrix}\hat{K}_{00}(x-x')\\ \hat{L}_{00}(x-x')\end{Bmatrix}\mathscr{A}_0(x)\,\mathscr{A}_0(x')\,, \qquad [10.2]$$

where

$$\mathscr{A}_0(x) = \begin{cases}1 & x\in V\\ 0 & x\notin V\,.\end{cases}$$

The quantity $\mathscr{A}_0(x)$ is analogous to the electrodynamic potential. Intuitively, these charge fluctuations are produced by the spontaneous fluctuations of pairs being created and again annihilated.

Now,

$$\langle Q_V^2\rangle_0 = \frac{e^2}{(2\pi)^4}\int \mathscr{A}_0(p)\,\mathscr{A}_0(-p)\begin{Bmatrix}\hat{K}_{00}(p)\\ \hat{L}_{00}(p)\end{Bmatrix}\mathrm{d}^4p\,,$$

where

$$\mathscr{A}_0(p) = \int \exp[-ipx]\,\mathscr{A}_0(x)\,\mathrm{d}^4x\,.$$

Note: This subject was first investigated by W. Heisenberg.[4]

If the space-time region is sharply defined, then the field intensities $\partial\mathscr{A}_0/\partial x^\mu$ have δ-function singularities, and the integrals diverge.

Modification: Let

$$\mathscr{A}_0(x) = \mathscr{A}_0'(t)\cdot\mathscr{A}_0''(\boldsymbol{x})\,,$$

[4] W. Heisenberg, *Ber. sächs. Akad. Wiss.*, p. 317 (1934).

where

$$\mathscr{A}_0'(t) = \begin{cases} 1 & |t| \leqslant T \\ 0 & |t| > T\,, \end{cases}$$

$$\mathscr{A}_0''(\boldsymbol{x}) = \begin{cases} 1 & 0 \leqslant |\boldsymbol{x}| \leqslant R \\ \exp[-\lambda(|\boldsymbol{x}| - R)] & |\boldsymbol{x}| \geqslant R, \quad \lambda = \dfrac{1}{b}\,. \end{cases}$$

Intuitively this means that a force flux is measured over a three-dimensional region with a corresponding weighting factor, instead of over a two-dimensional region:

$$\oint E_n \mathrm{d}f \to \int_{r>R} g(r) E_r \mathrm{d}^3x\,.$$

With this,

$$\mathscr{A}_0(p) = \lambda \frac{\sin p_0 T}{p_0} \cdot \frac{1}{p^2} \left[\frac{1}{p} \left(\frac{\lambda + Rp^2}{\lambda^2 + p^2} + \frac{2\lambda p^2}{(\lambda^2 + p^2)^2} \right) \sin pR + \left(\frac{1 - \lambda R}{\lambda^2 + p^2} + \frac{p^2 - \lambda^2}{(p^2 + \lambda^2)^2} \right) \cos pR \right].$$

Estimate: With $\lambda \gg 1/T \sim 1/R \gg m$, that is, $b \ll T \sim R \ll 1/m$, one obtains

$$\langle Q_V^2 \rangle_0 \sim e^2 R^2 \log \frac{R}{b} \qquad \text{(for both values of spin)}\,. \qquad [10.3]$$

(Heisenberg, on the other hand, considers the case $T \ll b \ll R \ll 1/m$.)

11. THE "CAUSAL" KERNELS $K_{\mu\nu}^c$ AND $L_{\mu\nu}^c$

We had

$$K_{\mu\nu}^c(p) = 4 \left(\frac{1}{2\pi} \right)^4 \int \frac{1}{k^2 + m^2 - i\mu^2} \cdot \frac{1}{(k-p)^2 + m^2 - i\mu^2} \cdot [-k_\mu (p-k)_\nu - k_\nu (p-k)_\mu - \delta_{\mu\nu}(-(p-k)_\lambda k_\lambda + m^2)]\, \mathrm{d}^4k\,, \qquad [11.1]$$

$$K_{\mu\nu}^c(p) = \hat{K}_{\mu\nu}(p) + i\overline{K}_{\mu\nu}(p) \text{ (formally!)}\,.$$

With

$$\frac{1}{ab}=\frac{1}{2}\int_{-1}^{+1}\frac{\mathrm{d}u}{\left(\frac{a+b}{2}+\frac{a-b}{2}u\right)^2} \quad \text{(after Feynman [5])}, \qquad [11.2]$$

we get

$$K^C_{\mu\nu}(p)=4\cdot\left(\frac{1}{2\pi}\right)^4\cdot\frac{1}{2}\int\mathrm{d}^4k\int_{-1}^{+1}\mathrm{d}u$$

$$\cdot\frac{-k_\mu(p-k)_\nu-k_\nu(p-k)_\mu-\delta_{\mu\nu}(-(p-k)_\lambda k_\lambda+m^2)}{[(k(k-p))+\frac{1}{2}p^2+m^2-i\mu^2+((kp)-\frac{1}{2}p^2)u]^2}.$$

We undertake to complete a square in the denominator:

$$k_\nu=K_\nu+\tfrac{1}{2}p_\nu(1-u),$$
$$k_\nu-p_\nu=K_\nu-\tfrac{1}{2}p_\nu(1+u).$$

Actually, this displacement is permitted only if the whole expression is regularized. Then,

denominator: $K^2+\frac{1}{4}p^2(1-u^2)+m^2-i\mu^2$,

numerator: $2K_\mu K_\nu-\frac{1}{2}p_\mu p_\nu(1-u^2)$
$-\delta_{\mu\nu}[K^2-\frac{1}{4}p^2(1-u^2)+m^2]$

plus terms linear in K_μ which do not contribute, as can be concluded from symmetry arguments. Rigorously, this is also only true upon regularization.

We now split up $K^C_{\mu\nu}$:

$$K^C_{\mu\nu}=(K^C_{\mu\nu})_I+(K^C_{\mu\nu})_{II},$$

where

$$(K^C_{\mu\nu})_I=4\cdot\frac{1}{2}\left(\frac{1}{2\pi}\right)^4\int_{-1}^{+1}\mathrm{d}u\int\mathrm{d}^4K$$

$$\cdot\frac{2K_\mu K_\nu-\delta_{\mu\nu}(K^2+\frac{1}{4}p^2(1-u^2)+m^2)}{[K^2+\frac{1}{4}p^2(1-u^2)+m^2-i\mu^2]^2},$$

$$=4\cdot\frac{1}{2}\left(\frac{1}{2\pi}\right)^4\int_{-1}^{+1}\mathrm{d}u\int\mathrm{d}^4K\left(\frac{2K_\mu K_\nu}{N^2}-\frac{\delta_{\mu\nu}}{N}\right),$$

[5] R. P. Feynman, *Phys. Rev.* **76**, 769 (1949); Appendix.

and

$$N \equiv K^2 + \tfrac{1}{4}p^2(1-u^2) + m^2 - i\mu^2 .$$

We have here neglected a term which contains μ^2 in the numerator—this is valid in the limit as $\mu \to 0$.

We will show that the term $(K^C_{\mu\nu})_I$, which is not gauge invariant, vanishes upon regularization (photon self-energy). The term $(K^C_{\mu\nu})_{II}$ is gauge invariant:

$$(K^C_{\mu\nu})_{II} = -(p_\mu p_\nu - \delta_{\mu\nu}p^2)\left(\frac{1}{2\pi}\right)^4 \int_{-1}^{+1}(1-u^2)\,\mathrm{d}u \int \frac{\mathrm{d}^4K}{N^2} .$$

A new grouping with the help of a partial integration with respect to u is

$$(K^C_{\mu\nu})_{II} = (K^C_{\mu\nu})_{IIa} + (K^C_{\mu\nu})_{IIb} ,$$

$$(K^C_{\mu\nu})_{IIa} = -(p_\mu p_\nu - \delta_{\mu\nu}p^2)\cdot\frac{4}{3}\left(\frac{1}{2\pi}\right)^4 \int \frac{\mathrm{d}^4K}{(K^2+m^2-i\mu^2)^2} . \quad [11.3]$$

The term $(K^C_{\mu\nu})_{IIa}$ is the self-charge. This is seen as follows. The current of the external field is

$$j^{\mu(a)}(x) = \frac{\partial F^{\text{ext}}_{\mu\nu}}{\partial x^\nu} = \frac{\partial^2 \mathscr{A}_\nu}{\partial x^\mu \partial x^\nu} - \Box \mathscr{A}_\mu .$$

In momentum space,

$$j^{\mu(a)}(x) \sim (p_\mu p_\nu - \delta_{\mu\nu}p^2)\,\mathscr{A}_\nu(p) .$$

However, the induced current is $K_{\mu\nu}(p)\mathscr{A}_\nu(p)$; therefore, if

$$K_{\mu\nu}(p) = \text{constant}\cdot(p_\mu p_\nu - \delta_{\mu\nu}p^2) ,$$

then this has the significance of a self-charge.

The remaining integral is regular:

$$(K^C_{\mu\nu})_{IIb} = +p^2(p_\mu p_\nu - \delta_{\mu\nu}p^2)$$

$$\cdot\left(\frac{1}{2\pi}\right)^4 \int_{-1}^{+1}\left(u^2 - \frac{u^4}{3}\right)\mathrm{d}u \int \frac{1}{N^3}\,\mathrm{d}^4K . \quad [11.4]$$

According to Feynman, the following is true. If $\mathrm{Im} L > 0$, then

$$\left(\frac{1}{2\pi}\right)^2 \int \frac{\mathrm{d}^4 K}{(K^2 - L)^3} = -\frac{i}{8L} . \qquad [11.5]$$

(For $\mathrm{Im} L < 0$, this becomes $+i/8L$.) Then,

$$(K^C_{\mu\nu})_{IIb} = +(p_\mu p_\nu - \delta_{\mu\nu} p^2) p^2 \cdot \frac{i}{8}\left(\frac{1}{2\pi}\right)^2 \int_{-1}^{+1} \frac{u^2 - u^4/3}{\frac{1}{4}p^2(1-u^2) + m^2 - i\mu^2} \mathrm{d}u , \qquad [11.6]$$

$$(\overline{K}_{\mu\nu})_{\mathrm{reg}} = +(p_\mu p_\nu - \delta_{\mu\nu} p^2) p^2 \cdot \frac{1}{8}\left(\frac{1}{2\pi}\right)^2 \mathscr{P}\int_{-1}^{+1} \frac{u^2 - u^4/3}{\frac{1}{4}p^2(1-u^2) + m^2} \mathrm{d}u . \qquad [11.7]$$

Analogously, using Eq. [11.2], we obtain

$$L^C_{\mu\nu}(p) = -\left(\frac{1}{2\pi}\right)^4 \int \mathrm{d}^4 k \, M_{\mu\nu} \cdot \left[(k^2 + m^2 - i\mu^2)((k-p)^2 + m^2 - i\mu^2)\right]^{-1}$$

$$= -\frac{1}{2}\left(\frac{1}{2\pi}\right)^4 \int \mathrm{d}^4 k \int_{-1}^{+1} \mathrm{d}u \, M_{\mu\nu} \cdot \left[k(k-p) + \tfrac{1}{2}p^2 + m^2 - i\mu^2 + (kp - \tfrac{1}{2}p^2)u\right]^{-2} ,$$

where

$$M_{\mu\nu} = -k_\mu(p-k)_\nu - k_\nu(p-k)_\mu + k_\mu k_\nu + (p-k)_\mu(p-k)_\nu - \delta_{\mu\nu}(k^2 + 2m^2 + (p-k)^2) .$$

With $k_\nu = K_\nu + \frac{1}{2} p_\nu(1-u)$, the numerator becomes

$$M_{\mu\nu} = 4K_\mu K_\nu + p_\mu p_\nu u^2 - \delta_{\mu\nu}(2K^2 + \tfrac{1}{2}p^2(1+u^2) + 2m^2) + \text{terms linear in } K_\mu ,$$

while the denominator is the same as in $K^C_{\mu\nu}$. Upon performing exactly the same calculation as before, one finds

$$L^C_{\mu\nu}(p) = -\tfrac{1}{2}(K^C_{\mu\nu}(p))_I + \tfrac{1}{4}(K^C_{\mu\nu}(p))_{IIa} + (L^C_{\mu\nu})_{\mathrm{reg}} . \qquad [11.8]$$

The nongauge-invariant terms can be compensated; the self-charge can never be compensated.

$$\left(L^{C}_{\mu\nu}(p)\right)_{\text{reg}} = -\frac{i}{16}(p_\mu p_\nu - \delta_{\mu\nu}\cdot p^2) \cdot p^2\left(\frac{1}{2\pi}\right)^2 \int_{-1}^{+1} \frac{\frac{1}{3}u^4}{\frac{1}{4}p^2(1-u^2)+m^2-i\mu^2}\,du\,, \qquad [11.9]$$

$$\left(\bar{L}_{\mu\nu}(p)\right)_{\text{reg}} = -\frac{1}{16}(p_\mu p_\nu - \delta_{\mu\nu}\cdot p^2) \cdot p^2\cdot\left(\frac{1}{2\pi}\right)^2 \mathscr{P}\int_{-1}^{+1} \frac{\frac{1}{3}u^4}{\frac{1}{4}p^2(1-u^2)+m^2}\,du\,. \qquad [11.10]$$

The real parts of K^C and L^C, that is, $\hat{K}$ and $\hat{L}$, are found as residues:

Real part $= i\pi \times$ residue (Jost and Luttinger [6]) .

There is a contribution only if the zero of the denominator lies between zero and one; that is, if

$$0 < \frac{p^2+4m^2}{p^2} < 1\,,$$

or if

$$p^2 < -4m^2\,.$$

In this case, for spin-$\frac{1}{2}$ particles, the residue is at $u^2 = (p^2+4m^2)/p^2$:

$$\frac{u^2-\frac{1}{3}u^4}{-\frac{1}{2}p^2u} = -\frac{2}{p^2}u\left(1-\frac{u^2}{3}\right) = -\frac{2}{p^2}\cdot\frac{2p^2-4m^2}{3p^2}\sqrt{\frac{p^2+4m^2}{p^2}}\,.$$

Thus,

$$\hat{K}_{\mu\nu}(p) = \begin{cases} (p_\mu p_\nu - \delta_{\mu\nu}p^2)\,\dfrac{\pi}{2\pi^2}\cdot\dfrac{p^2-2m^2}{3p^2}\sqrt{\dfrac{p^2+4m^2}{p^2}}\,, & p^2 < -4m^2 \\ 0\,, & \text{otherwise}\,, \end{cases} \qquad [11.11]$$

[6] R. JOST and J. M. LUTTINGER, *Helv. Phys. Acta* **23**, 201 (1950).

$$\hat{L}_{\mu\nu}(p) = \begin{cases} \dfrac{1}{16}(p_\mu p_\nu - \delta_{\mu\nu}p^2)\left(\dfrac{1}{2\pi}\right)^2 \cdot \dfrac{\pi}{3}\left(\dfrac{p^2+4m^2}{p^2}\right), \\ \qquad\qquad\qquad\qquad p^2 < -4m^2 \\ 0\,, \quad \text{otherwise}\,, \end{cases} \qquad [11.12]$$

as before.

Discussion of the singular terms

With

$$N \equiv k^2 + \tfrac{1}{2}p^2(1-u^2) + m^2 - i\mu^2\,,$$

we have (see p. 43)

$$(K^C_{\mu\nu})_I = 2\cdot\left(\frac{1}{2\pi}\right)^4 \int_{-1}^{+1} du \int \left(\frac{2k_\mu k_\nu}{N^2} - \frac{\delta_{\mu\nu}}{N}\right) d^4k\,, \qquad [11.13]$$

and

$$(K^C_{\mu\nu})_{IIa} = -(p_\mu p_\nu - \delta_{\mu\nu}p^2)\left(\frac{1}{2\pi}\right)^4 \cdot \frac{4}{3}\int \frac{d^4k}{(k^2+m^2-i\mu^2)^2}\,. \qquad [11.14]$$

In this form the integrals are indeterminate. We regularize:

$$(\widetilde{K^C_{\mu\nu}})_I \equiv \sum_{i=0}^{N} C_i K_{\mu\nu}(p; M_i)_I\,; \qquad C_0 = 1,\ M_0 = m\,,$$

$$(\widetilde{K^C_{\mu\nu}})_{IIa} \equiv \sum_{i=0}^{N} C_i K_{\mu\nu}(p; M_i)_{IIa}\,.$$

For the integrals to be finite it suffices that

for I: $\sum_{i=0}^{N} C_i = 0$; $\sum_{i=0}^{N} C_i M_i^2 = 0$, so that the integral vanishes;

for IIa: $\sum_{i=0}^{N} C_i = 0$, and the integral is then determined by $\sum_{i=1}^{N} C_i \log(M_i/m)$.

1. The self-charge, $(K^C_{\mu\nu})_{IIa}$: One auxiliary mass $M_1 \equiv M$

suffices. With $\alpha \equiv e^2/4\pi \simeq 1/137$, one finds

$$\frac{\delta e}{e} = \frac{\alpha}{3\pi} \log \frac{m^2}{M^2} = \frac{2\alpha}{3\pi} \log \frac{m}{M} < 0 \qquad \text{(Schwinger [7])} . \qquad [11.15]$$

Note: In the Feynman-Dyson formalism, in which the definition of δe is given on the basis of scattering processes, δe is only half as large. (Interpretation?) [A 2].

We now carry out the calculation. We had Eq. [11.5],

$$\left(\frac{1}{2\pi}\right)^2 \int \frac{d^4k}{(k^2 - L)^3} = \frac{-i}{8L} \qquad (\operatorname{Im} L > 0) .$$

Thus,

$$\frac{1}{2}\left(\frac{1}{2\pi}\right)^2 \int \left[\frac{1}{(k^2 - L_2)^2} - \frac{1}{(k^2 - L_1)^2}\right] d^4 k = \frac{-i}{8} \log \frac{L_2}{L_1} . \qquad [11.16]$$

This corresponds to the case with one auxiliary mass. More generally,

$$\sum_{i=0}^{N-1} C_i \left(\frac{1}{2\pi}\right)^2 \frac{1}{2} \int \left[\frac{1}{(k^2 - L_N)^2} - \frac{1}{(k^2 - L_i)^2}\right] d^4k$$

$$= -\frac{i}{8} \sum_{i=0}^{N-1} C_i \log \frac{L_N}{L_i} .$$

We have $\sum_{i=0}^{N-1} C_i = -C_N$, since $\sum_{i=0}^{N} C_i = 0$. Therefore, corresponding to Eq. [11.16], we get

$$-\left(\frac{1}{2\pi}\right)^2 \cdot \frac{1}{2} \int \left[\sum_{i=0}^{N} \frac{C_i}{(k^2 - L_i)^2}\right] d^4k = \frac{i}{8} \sum_{i=0}^{N} C_i \log L_i . \qquad [11.17]$$

With this, Schwinger's value follows.

2. The photon self-energy: We have

$$(K^C_{\mu\nu})_I = 2 \cdot \left(\frac{1}{2\pi}\right)^4 \int_{-1}^{+1} du \int \left[\frac{2k_\mu k_\nu}{(k^2 - L)^2} - \delta_{\mu\nu} \frac{1}{k^2 - L}\right] d^4k ,$$

[7] J. SCHWINGER, *Phys. Rev.* **75**, 651 (1949).

where

$$L = -\tfrac{1}{4}p^2(1-u^2) - m^2 + i\mu^2 .$$

It is to be shown that this vanishes upon regularization. First, we have

$$\int \frac{k_\mu k_\nu}{(k^2-L)^4}\,\mathrm{d}^4k = \delta_{\mu\nu}\cdot\frac{1}{4}\int\frac{k^2}{(k^2-L)^4}\,\mathrm{d}^4k$$

$$= \frac{1}{4}\,\delta_{\mu\nu}\left\{\int\frac{\mathrm{d}^4k}{(k^2-L)^3} + L\int\frac{\mathrm{d}^4k}{(k^2-L)^4}\right\}.$$

From

$$\left(\frac{1}{2\pi}\right)^2\int(k^2-L)^{-3}\,\mathrm{d}^4k = \frac{-i}{8L} \qquad (\mathrm{Im}\,L>0)\,,$$

it follows, by differentiation with respect to L, that

$$3\left(\frac{1}{2\pi}\right)^2\int\frac{\mathrm{d}^4k}{(k^2-L)^4} = +\frac{i}{8L^2}\,.$$

Thus,

$$\int\frac{k_\mu k_\nu}{(k^2-L)^4}\,\mathrm{d}^4k = \frac{1}{6}\,\delta_{\mu\nu}\int\frac{\mathrm{d}^4k}{(k^2-L)^3}\,,$$

or

$$\int\left[\frac{k_\mu k_\nu}{(k^2-L)^4} - \frac{1}{6}\,\delta_{\mu\nu}\,\frac{1}{(k^2-L)^3}\right]\mathrm{d}^4k = 0\,.$$

If this is integrated twice with respect to L, we obtain exactly what we want: Multiplying by $\mathrm{d}L$ and integrating between L_1 and L_2,

$$\frac{1}{3}\int\left[\frac{1}{2}\,\frac{k_\mu k_\nu}{(k^2-L)^3} - \delta_{\mu\nu}\cdot\frac{1}{4}\,\frac{1}{(k^2-L)^2}\right]\mathrm{d}^4k\,\Bigg|_{L_1}^{L_2} = 0\,.$$

(This is true if the k integration is carried out last.) That is, if $\sum_i C_i = 0$, then we have

$$\int\sum_i C_i f(L_i, k)\,\mathrm{d}^4k = 0\,,$$

where

$$f(L_i, k) \equiv \frac{k_\mu k_\nu}{(k^2-L_i)^3} - \delta_{\mu\nu}\cdot\frac{1}{4}\,\frac{1}{(k^2-L_i)^2}\,.$$

Let

$$F(L, k) \equiv \frac{1}{2} \frac{k_\mu k_\nu}{(k^2 - L)^2} - \frac{1}{4} \delta_{\mu\nu} \frac{1}{k^2 - L}.$$

Then,

$$f = \frac{\partial F}{\partial L}.$$

One can, now, quite generally, say the following. Let

$$\int [f(k, L) - f(k, L_1)] \, \mathrm{d}^4 k = 0 .$$

Then,

$$\int \mathrm{d}^4 k \int_{L_1}^{L_2} [f(k, L) - f(k, L_1)] \, \mathrm{d}L = 0 ,$$

or

$$\int [F(k, L_2) - F(k, L_1) - (L_2 - L_1) f(k, L_1)] \, \mathrm{d}^4 k = 0 .$$

If $\sum_i C_i = 0$, then, more generally,

$$\int \sum_i C_i [F(k, L_i) - L_i f(k, L_1)] \, \mathrm{d}^4 k = 0 .$$

If, in addition, $\sum_i C_i L_i = 0$, then

$$\int \sum_i C_i F(k, L_i) \, \mathrm{d}^4 k = 0 . \qquad \text{Q.E.D.}$$

This method of integrating over the masses works quite generally.

Now, we have

$$L_i \equiv -\tfrac{1}{4} p^2 (1 - u^2) + i\mu^2 - M_i^2 .$$

Thus, $\sum_i C_i = 0$ and $\sum_i C_i L_i = 0$ are equivalent to $\sum_i C_i = 0$ and $\sum_i C_i M_i^2 = 0$. Hence it is shown that $(K^C_{\mu\nu})_I$ vanishes when regularized with $\sum_i C_i = 0$ and $\sum_i C_i M_i^2 = 0$.

12. IMPOSSIBILITY OF CANCELING THE SELF-CHARGE [8]

We assert that if e_0 is the bare charge, e the physical charge, $\alpha = e^2/4\pi \simeq 1/137$ the fine-structure constant, and

$$\frac{e}{e_0} \equiv \frac{1}{1+F(\alpha)}, \qquad [12.1]$$

then

$$\left.\begin{aligned} F(\alpha) > 0\,; \\ 0 < \frac{e}{e_0} < 1 \end{aligned}\right\} \qquad [12.2]$$

that is,

is true.

Note: 1. We know the first approximation (in α) of this assertion: $\delta e = -eF(\alpha)$ is, in first approximation, negative.

2. This question is unrelated to Lorentz invariance; the canonical formalism can be employed. Now, in the Heisenberg representation, with $\dot{F} \equiv \partial F/\partial t$,

$$[\Phi_\mu(\boldsymbol{x}, t), \dot{\Phi}_\nu(\boldsymbol{x}', t)] = i\delta_{\mu\nu}\delta^3(\boldsymbol{x} - \boldsymbol{x}')\,. \qquad [12.3]$$

The reason why this is rigorously true is that the interaction energy is dependent only upon the potentials and not on the field intensities. With a Pauli term [A-3] it would no longer be valid.

Let Φ_μ be a rigorous solution of the field equations including interactions:

$$\Box\Phi_\mu = -j^\mu\,.$$

Let $\Phi_\mu^{(0)} = \lim_{\alpha\to 0}\Phi_\mu$; that is, $\Box\Phi_\mu^{(0)} = 0$. Then, Eq. [12.3] is valid for $\Phi_\mu^{(0)}$ as well as for Φ_μ. Now, one can write

$$\Phi_\mu = \gamma\Phi_\mu^{(0)} + \Phi_\mu^{(1)}, \qquad [12.4]$$

where $\Phi_\mu^{(1)}$ changes either (*a*) the number of pairs of material particles, (*b*) the number of photons by three or more, or (*c*) both.

[8] J. SCHWINGER, *Phys. Rev.* **76**, 790 (1949); Appendix.

Then,

$$\left.\begin{aligned}\langle \Phi_\mu^{(0)}(\boldsymbol{x},t)\,\dot{\Phi}_\nu^{(1)}(\boldsymbol{x}',t')\rangle_0 &= 0\\ \langle \dot{\Phi}_\mu^{(0)}(\boldsymbol{x},t)\,\Phi_\nu^{(1)}(\boldsymbol{x}',t')\rangle_0 &= 0\end{aligned}\right\}. \qquad [12.5]$$

From this, with $\gamma = e/e_0$,

$$\langle[\Phi_\mu(\boldsymbol{x},t),\,\dot{\Phi}_\mu(\boldsymbol{x}',t)]\rangle_0 = \gamma^2\langle[\Phi_\mu^{(0)}(\boldsymbol{x},t),\,\dot{\Phi}_\mu^{(0)}(\boldsymbol{x}',t)]\rangle_0 + \langle[\Phi_\mu^{(1)}(\boldsymbol{x},t),\,\dot{\Phi}_\mu^{(1)}(\boldsymbol{x}',t)]\rangle_0 \quad (\text{no sum over } \mu)\,,$$

which implies

$$(1-\gamma^2)\delta^3(\boldsymbol{x}-\boldsymbol{x}') = -i\langle[\Phi_\mu^{(1)}(\boldsymbol{x},t),\,\dot{\Phi}_\mu^{(1)}(\boldsymbol{x}',t)]\rangle_0\,.$$

For one normal mode (r) in a box,

$$1-\gamma^2 = -i\langle[\Phi_\mu^{(1)}(k_r,t),\,\dot{\Phi}_\mu^{(1)}(k_r,t)]\rangle_0\,. \qquad [12.6]$$

Assertion: The right side is positive. We have, now,

$$\dot{\Phi}^{(1)} = +i[H,\Phi^{(1)}]\,.$$

Thus,

$$1-\gamma^2 = +\langle[\Phi^{(1)},[H,\Phi^{(1)}]]\rangle_0\,,$$

$$1-\gamma^2 = +\langle\Psi_0|2\Phi^{(1)}H\Phi^{(1)} - \Phi^{(1)}\Phi^{(1)}H - H\Phi^{(1)}\Phi^{(1)}|\Psi_0\rangle\,.$$

With $H\Psi_0 = E_0\Psi_0$ and the Hermiticity of H,

$$1-\gamma^2 = 2(\Phi^{(1)}\Psi_0,\,H\Phi^{(1)}\Psi_0) - 2(\Phi^{(1)}\Psi_0,\,E_0\Phi^{(1)}\Psi_0)\,, \qquad [12.7]$$

$$1-\gamma^2 = 2\big(\Phi^{(1)}\Psi_0,\,(H-E_0)\,\Phi^{(1)}\Psi_0\big)\,.$$

If Ψ_0 is the state of lowest energy, then this is positive.

Q.E.D.

Note: 1. The Hermiticity of H is essential.

2. This proof is not valid for theories with negative energies (Bopp and Stueckelberg [9]), since there $(\Psi^*(H-E_0)\Psi) \ngeq 0$ for all Ψ.

[9] F. Bopp, *Ann. Physik* **38**, 345 (1940); E. C. G. Stueckelberg, *Nature* **144**, 118 (1939).

Chapter 3. Quantization of Free Fields: Spin 0 and ½, Quantum Electrodynamics

13. THE INVARIANT FUNCTIONS [A-4]

The homogeneous wave equation is

$$(\square - m^2)\Delta = 0\,. \qquad [13.1]$$

We denote the solution of this equation corresponding to the initial conditions

$$\Delta(\boldsymbol{x}, 0) = 0$$

and

$$\left(\frac{\partial \Delta}{\partial t}\right)_{\boldsymbol{x},0} = -\,\delta^3(\boldsymbol{x})$$

at $t=0$ by $\Delta(x) \equiv \Delta(\boldsymbol{x}, t)$. Its Fourier representation is

$$\Delta(x) = -\left(\frac{1}{2\pi}\right)^3 \int \frac{\sin(\omega x_0)}{\omega} \exp[i\boldsymbol{k}\cdot\boldsymbol{x}]\,\mathrm{d}^3k \qquad (\omega \equiv \overset{+}{\sqrt{\boldsymbol{k}^2 + m^2}})\,,$$

as one readily recognizes when one considers that

$$\delta^3(\boldsymbol{x}) = \left(\frac{1}{2\pi}\right)^3 \int \exp[i\boldsymbol{k}\cdot\boldsymbol{x}]\,\mathrm{d}^3k\,.$$

Thus,

$$\Delta(x) = -i\left(\frac{1}{2\pi}\right)^3 \int \exp[i(kx)]\,\varepsilon(k)\,\delta(k^2 + m^2)\,\mathrm{d}^4k\,, \qquad [13.2]$$

where

$$\varepsilon(x) \equiv \varepsilon(t) = \begin{cases} +1 & (t>0) \\ -1 & (t<0) \\ \text{undetermined for } t=0 \,, \end{cases}$$

and

$$\frac{\partial \varepsilon}{\partial t} = 2\delta(t) \,.$$

Thus,

$$\Delta(\boldsymbol{x}, -t) = -\Delta(\boldsymbol{x}, t) \,, \qquad \Delta(-\boldsymbol{x}, t) = +\Delta(\boldsymbol{x}, t) \,,$$

and

$$\Delta(-x) = -\Delta(x) \,. \tag{13.3}$$

We can define an additional invariant solution of Eq. [13.1] by means of its Fourier decomposition:

$$\Delta^1(x) = \left(\frac{1}{2\pi}\right)^3 \int \exp[i(kx)]\delta(k^2+m^2)\, \mathrm{d}^4k \,, \tag{13.4}$$

or

$$\Delta^1(x) = \left(\frac{1}{2\pi}\right)^3 \int \frac{\cos \omega x_0}{\omega} \exp[i\boldsymbol{k}\cdot\boldsymbol{x}]\, \mathrm{d}^3k \,.$$

The relation to the advanced and retarded potentials is given by the functions

$$\Delta^{\text{ret}}(x) = \begin{cases} -\Delta(x) & (t>0) \\ 0 & (t<0) \end{cases} : \quad \Delta^{\text{ret}}(x) = -\frac{1+\varepsilon(x)}{2}\Delta(x) \,,$$

$$\Delta^{\text{adv}}(x) = \begin{cases} 0 & (t>0) \\ \Delta(x) & (t<0) \end{cases} : \quad \Delta^{\text{adv}}(x) = +\frac{1-\varepsilon(x)}{2}\Delta(x) \,.$$

We have

$$(\Box - m^2)\Delta^{\text{ret}}(x) = -\delta^4(x) \,,$$
$$(\Box - m^2)\Delta^{\text{adv}}(x) = -\delta^4(x) \,,$$

and

$$\left.\begin{aligned} &\Delta = \Delta^{\text{adv}} - \Delta^{\text{ret}} \\ &\bar{\Delta} = \tfrac{1}{2}(\Delta^{\text{adv}} + \Delta^{\text{ret}}) : \quad \bar{\Delta}(x) = -\tfrac{1}{2}\varepsilon(x)\Delta(x) \end{aligned}\right\} . \tag{13.5}$$

The Fourier representation of $\bar{\Delta}(x) = -\frac{1}{2}\varepsilon(x)\Delta(x)$ is

$$\bar{\Delta}(x) = \left(\frac{1}{2\pi}\right)^4 \mathscr{P}\int \frac{\exp[i(kx)]}{k^2+m^2}\, \mathrm{d}^4k \,, \tag{13.6}$$

where $\mathscr{P}$ denotes the principal value with respect to the k_0 integration. One sees immediately that

$$(\square - m^2)\bar{\Delta}(x) = - \delta^4(x)$$

is, in fact, satisfied. Moreover, the function has the correct symmetry property. This suffices to determine $\bar{\Delta}$.

The decomposition according to positive and negative frequencies occurs by means of

$$\left.\begin{aligned} \Delta^+ &= \tfrac{1}{2}(\Delta - i\Delta^1) \\ \Delta^- &= \tfrac{1}{2}(\Delta + i\Delta^1) \end{aligned}\right\} , \qquad [13.7]$$

as is recognized immediately from the representations in three-dimensional momentum space.

a. Representation by means of paths in the complex k_0-plane

$$\begin{aligned} \Delta^{\text{ret}}(x) &\equiv \bar{\Delta}(x) - \tfrac{1}{2}\Delta(x) \\ &= \left(\frac{1}{2\pi}\right)^4 \int \left\{ \mathscr{P} \frac{1}{k^2 + m^2} + i\pi\delta(k^2 + m^2)\varepsilon(k) \right\} \exp[i(kx)] \, d^4k , \end{aligned}$$

$$\left.\begin{aligned} \Delta^{\text{ret}} &= \left(\frac{1}{2\pi}\right)^4 \int_{\Gamma_+} \frac{\exp[i(kx)]}{k^2 + m^2} \, d^4k \\ \Delta^{\text{adv}} &= \left(\frac{1}{2\pi}\right)^4 \int_{\Gamma_-} \frac{\exp[i(kx)]}{k^2 + m^2} \, d^4k \end{aligned}\right| . \qquad [13.8]$$

(See Fig. 13.1.)

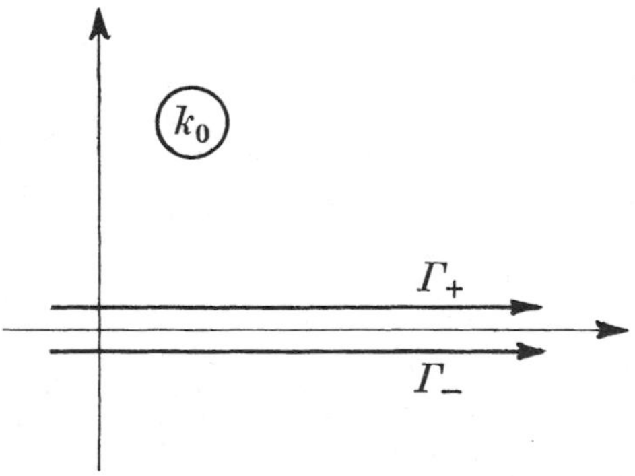

Figure 13.1

The Δ^c function is defined by

$$\Delta^c \equiv -2i\bar{\Delta} + \Delta^1, \qquad [13.9]$$

$$(\Box - m^2)\Delta^c = +2i\delta^4(x) .$$

We have

$$\Delta^c(x) = \frac{-2i}{(2\pi)^4}\int \exp[i(kx)]\left[\mathscr{P}\frac{1}{k^2+m^2} + i\pi\delta(k^2+m^2)\right]\mathrm{d}^4k .$$

From

$$\left.\begin{aligned} \delta_+(z) &\equiv \frac{1}{2}\left[\delta(z) + \mathscr{P}\frac{1}{i\pi z}\right] = \frac{1}{2\pi}\int_0^\infty \exp[-i\nu z]\mathrm{d}\nu \\ \delta_-(z) &\equiv \frac{1}{2}\left[\delta(z) - \mathscr{P}\frac{1}{i\pi z}\right] = \frac{1}{2\pi}\int_{-\infty}^0 \exp[-i\nu z]\mathrm{d}\nu \end{aligned}\right\}, \qquad [13.10]$$

there follows

$$\Delta^c(x) = \frac{4\pi}{(2\pi)^4}\int \exp[i(kx)]\delta_+(k^2+m^2)\,\mathrm{d}^4k . \qquad [13.11]$$

In terms of a path in the complex plane (see Fig. 13.2),

$$\Delta^c(x) = \frac{-2i}{(2\pi)^4}\int_C \frac{\exp[i(kx)]}{k^2+m^2}\,\mathrm{d}^4k . \qquad [13.12]$$

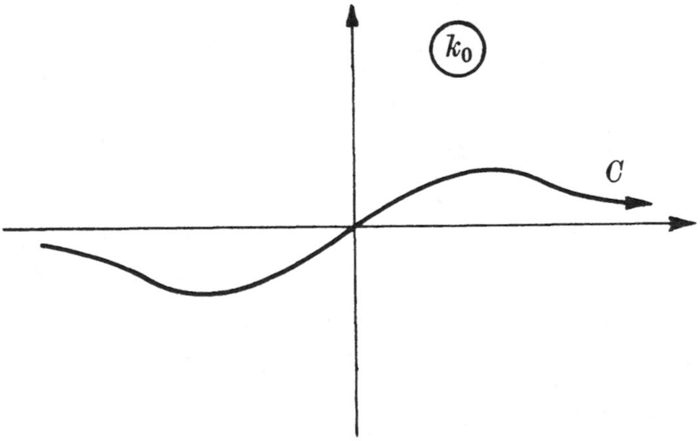

Figure 13.2

b. Decomposition of Δ^C in terms of positive and negative frequencies (Fierz [1])

$$\left.\begin{aligned}(\Delta^C)^+ &= \frac{-2i}{(2\pi)^4}\int_{C_+} \frac{\exp[i(kx)]}{k^2+m^2}\,\mathrm{d}^4k \\ (\Delta^C)^- &= \frac{-2i}{(2\pi)^4}\int_{C_-} \frac{\exp[i(kx)]}{k^2+m^2}\,\mathrm{d}^4k\end{aligned}\right\}. \qquad [13.13]$$

(See Fig. 13.3.) Then,

$$\left.\begin{aligned}(\Delta^C)^+ &= -2i(\Delta^{\mathrm{ret}})^+ \\ (\Delta^C)^- &= -2i(\Delta^{\mathrm{adv}})^-\end{aligned}\right\} \qquad [13.14]$$

(Fierz's decomposition of the Δ^C functions; see Section 27),

$$\Delta^C = -2i[(\Delta^{\mathrm{ret}})^+ + (\Delta^{\mathrm{adv}})^-]. \qquad [13.15]$$

Corresponding decomposition of the sources:

$$\left.\begin{aligned}(\Box - m^2)(\Delta^C)^+ &= +2i\delta^3(x)\delta_+(t) \\ (\Box - m^2)(\Delta^C)^- &= +2i\delta^3(x)\delta_-(t)\end{aligned}\right\}. \qquad [13.16]$$

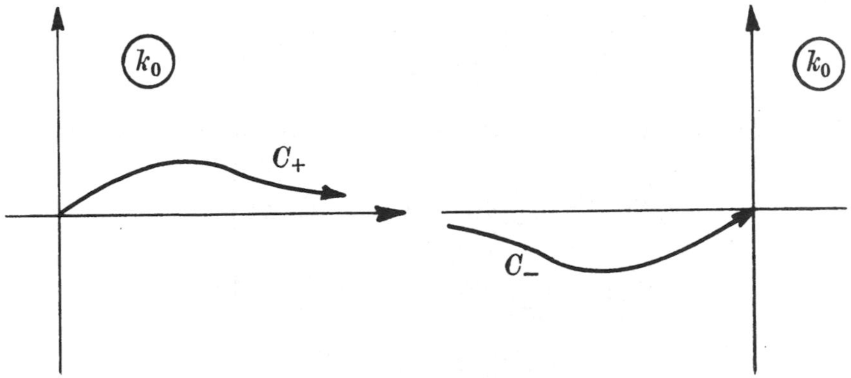

Figure 13.3

That is, Fierz's decomposition of Δ^C corresponds to the decomposition of the sources in terms of positive and negative frequencies.

[1] M. Fierz, *Helv. Phys. Acta* **23**, 731 (1950).

c. Feynman's decomposition of Δ^c

This is a decomposition in terms of positive and negative times:

$$\Delta^c(x) = i\varepsilon(x)\Delta(x) + \Delta^1(x) \ ,$$

so that for

$$\begin{aligned} t>0, \quad \Delta^c &= \ \ i\Delta + \Delta^1 = \ \ 2i\Delta^+, \\ t<0, \quad \Delta^c &= -i\Delta + \Delta^1 = -2i\Delta^-, \end{aligned}$$

or

$$\Delta^c = 2i(\Delta^+)^{\text{ret}} - 2i(\Delta^-)^{\text{adv}} .$$

Note: This is not a frequency decomposition according to the sign of the frequency:

$$(\Delta^+)^{\text{ret}} \neq (\Delta^{\text{ret}})^+ .$$

d. Addendum. The special case of vanishing mass, $m=0$

$$D(x) = \Delta(x)|_{m=0} \ , \qquad \text{etc.} \ ,$$

$$\left.\begin{aligned} \Box D(x) &= 0 \\ \Box D^1(x) &= 0 \\ \Box \bar{D}(x) &= -\delta^4(x) \end{aligned}\right\} .$$

Then, explicit expressions for the D functions can be given:

$$\left.\begin{aligned} D^{\text{ret}}(x) &= \frac{1}{4\pi r}\,\delta(t-r)\,; \qquad r \equiv |\boldsymbol{x}| \\ D^{\text{adv}}(x) &= \frac{1}{4\pi r}\,\delta(t+r) \end{aligned}\right\} , \qquad [13.17]$$

$$D(x) = D^{\text{adv}}(x) - D^{\text{ret}}(x) \ ,$$

$$\bar{D}(x) = \tfrac{1}{2}\left(D^{\text{adv}}(x) + D^{\text{ret}}(x)\right) .$$

Thus,

$$\left.\begin{aligned} D(x) &= \frac{1}{4\pi r}\,[\delta(t+r) - \delta(t-r)] \\ \bar{D}(x) &= \frac{1}{8\pi r}\,[\delta(t+r) + \delta(t-r)] \end{aligned}\right\} , \qquad [13.18]$$

and

$$(D^{\text{ret}})^{+}(x) = \frac{1}{4\pi r}\,\delta_{+}(t-r)\,,$$

$$(D^{\text{adv}})^{-}(x) = \frac{1}{4\pi r}\,\delta_{-}(t+r)\,,$$

$$D^{C} = -\,2i[(D^{\text{ret}})^{+} + (D^{\text{adv}})^{-}]\,,$$

$$= \frac{-\,2i}{8\pi r}\left[\delta(t-r) + \mathscr{P}\frac{1}{i\pi(t-r)} + \delta(t+r) - \mathscr{P}\frac{1}{i\pi(t+r)}\right]$$

$$= -\,2i\,\bar{D} - \frac{2i}{8\pi r}\mathscr{P}\frac{2r}{i\pi(t^2-r^2)}\,,$$

$$D^{C} = -\,2i\bar{D} + \frac{1}{2\pi^2}\mathscr{P}\frac{1}{r^2-t^2}\,,$$

$$D^{C} = -\,2i\bar{D} + D^1\,,$$

where

$$D^1(x) = \mathscr{P}\frac{1}{2\pi^2(r^2-t^2)}\,. \qquad [13.19]$$

Note: For $m \neq 0$, these same expressions represent the parts of the functions in question having the stronger singularity; in that case, however, there are also additional terms with weaker singularities (logarithms for the case of Δ^1 and discontinuities for Δ and $\bar{\Delta}$). These additional terms can also be written down explicitly (cf. Schwinger [2]).

14. QUANTIZATION OF FORCE-FREE, UNCHARGED, SPIN-0 FIELDS [A-4]

Let $\Phi(x)$ be a real scalar field:

$$(\Box - m^2)\,\Phi(x) = 0\,. \qquad [14.1]$$

In a large volume V,

$$\Phi(x) = \frac{1}{\sqrt{V}}\sum_{\substack{k\\k_0>0}}\frac{1}{\sqrt{2\omega}}\,[A_k \exp[i(kx)] + A_k^* \exp[-\,i(kx)]]\,. \qquad [14.2]$$

[2] J. Schwinger, *Phys. Rev.* **75**, 651 (1949).

Here, the sum is to be extended over all values of $\boldsymbol{k}$; the frequency is always to be chosen positive:

$$k_0 = +\omega = +\sqrt{\boldsymbol{k}^2 + m^2}\,,$$

$$(kx) \equiv \boldsymbol{k}\cdot\boldsymbol{x} - \omega t\,.$$

In the transition to $V \to \infty$, one must write

$$\frac{1}{V}\sum_{\boldsymbol{k}} \ldots \to \left(\frac{1}{2\pi}\right)^3 \int \ldots \, \mathrm{d}^3k\,, \qquad A_k \to A(\boldsymbol{k})\,.$$

The commutation relations

$$[A_k, A_l^*] = \delta_{kl}$$

hold. Thus,

$$A_k^* A_k = N_k \qquad \text{(with eigenvalues of } 0, 1, 2, \ldots)\,,$$

and A_k is an absorption or annihilation operator while A_k^* is an emission or creation operator:

$$\left.\begin{aligned} A\Psi(N) &= \sqrt{N}\,\Psi(N-1) \\ A^*\Psi(N) &= \sqrt{N+1}\,\Psi(N+1) \end{aligned}\right\}.$$

For the vacuum,

$$\left.\begin{aligned} \langle A^*A\rangle_0 &= 0 \\ \langle AA^*\rangle_0 &= 1 \end{aligned}\right\},$$

so that

$$\left.\begin{aligned} [\Phi(x), \Phi(x')] &= i\Delta(x-x') \\ \langle\{\Phi(x), \Phi(x')\}\rangle_0 &= \Delta^{(1)}(x-x') \end{aligned}\right\}, \qquad [14.3]$$

using the notation of Section 9.

Canonical Formalism

Using the Lagrangian function

$$L = \int \mathscr{L} \mathrm{d}^3x\,, \qquad [14.4]$$

where

$$-\mathscr{L} = \frac{1}{2}\left[\frac{\partial\Phi}{\partial x^\nu}\frac{\partial\Phi}{\partial x^\nu} + m^2\Phi^2\right], \qquad [14.5]$$

the field equations can be obtained from a variational principle:

$$\delta\int L\,\mathrm{d}t = 0\,. \qquad [14.6]$$

Equation [14.5] can be written

$$2\mathscr{L} = \left(\frac{\partial\Phi}{\partial t}\right)^2 - \left(\frac{\partial\Phi}{\partial \boldsymbol{x}}\right)^2 - m^2\Phi^2\,. \qquad [14.7]$$

One defines the canonically conjugate momentum as in mechanics,

$$\pi(x) \equiv \left(\frac{\delta L}{\delta(\partial\Phi/\partial t)}\right)_\Phi = \frac{\partial\mathscr{L}}{\partial(\partial\Phi/\partial t)} = \frac{\partial\Phi}{\partial t}\,, \qquad [14.8]$$

and forms a Hamiltonian function

$$H = \int \mathscr{H}\,\mathrm{d}^3x\,, \qquad [14.9]$$

where

$$\mathscr{H} = \pi\frac{\partial\Phi}{\partial t} - \mathscr{L}\,. \qquad [14.10]$$

In doing this, the order of the factors may have to be respected. We have then

$$\mathscr{H} = \frac{1}{2}\left\{\pi^2(x) + \left(\frac{\partial\Phi}{\partial\boldsymbol{x}}\right)^2 + m^2\Phi^2(x)\right\}\,. \qquad [14.11]$$

Further utilizing the analogy to mechanics, we will demand the commutation relation

$$i[\pi(\boldsymbol{x}, t), \Phi(\boldsymbol{x}', t)] = \delta^3(\boldsymbol{x} - \boldsymbol{x}')\,, \qquad [14.12]$$

and, naturally,

$$[\pi(\boldsymbol{x}, t), \pi(\boldsymbol{x}', t)] = [\Phi(\boldsymbol{x}, t), \Phi(\boldsymbol{x}', t)] = 0\,, \qquad [14.13]$$

which is explicitly valid only for equal times. It is immediately seen that these commutation relations agree with the invariants of Eq. [14.3] if there one sets $t = t'$ and notes that $\pi = \partial\Phi/\partial t$, because

$$\Delta(\boldsymbol{x}-\boldsymbol{x}', 0) = 0\,, \qquad \frac{\partial\Delta}{\partial t}(\boldsymbol{x}-\boldsymbol{x}', 0) = -\,\delta^3(\boldsymbol{x}-\boldsymbol{x}')\,.$$

Conversely, the invariants also follow from the canonical commutation relations if the field equations are employed. In the canonical formalism, these equations result from the relation

$$\frac{\partial F}{\partial t} = i[H, F]: \tag{14.14}$$

$$\frac{\partial\Phi}{\partial t} = i[H, \Phi] = \pi\,, \qquad \frac{\partial\pi}{\partial t} = i[H, \pi] = \Delta\Phi - m^2\Phi\,.$$

15. QUANTUM ELECTRODYNAMICS IN VACUUM

For a real vector field Φ_μ, the field equations are

$$\Box\Phi_\mu = 0 \qquad \text{(zero mass)}\,. \tag{15.1}$$

In addition, we also demand the auxiliary condition

$$\frac{\partial\Phi_\mu}{\partial x^\mu} = 0 \tag{15.2}$$

in order that the field intensities

$$F_{\mu\nu} = \frac{\partial\Phi_\nu}{\partial x^\mu} - \frac{\partial\Phi_\mu}{\partial x^\nu} \tag{15.3}$$

satisfy the Maxwell equations

$$\frac{\partial F_{\mu\nu}}{\partial x^\nu} = 0\,. \tag{15.4}$$

For quantization, it is useful once again, in analogy to the scalar field, to require that

$$[\Phi_\mu(x), \Phi_\nu(x')] = i\delta_{\mu\nu}D(x-x')\,. \tag{15.5}$$

Here, a peculiarity appears, namely,

$$\left[\frac{\partial \Phi_\mu(x)}{\partial x^\mu}, \Phi_\nu(x')\right] = i\frac{\partial}{\partial x^\nu} D(x-x') \neq 0 . \qquad [15.6]$$

That is, the auxiliary condition does not commute with the potentials. However, it commutes with the field intensities,

$$\left[\frac{\partial \Phi_\varrho(x)}{\partial x^\varrho}, F_{\mu\nu}(x')\right] = 0 , \qquad [15.7]$$

and with itself,

$$\left[\frac{\partial \Phi_\mu(x)}{\partial x^\mu}, \frac{\partial \Phi_\nu(x')}{\partial x'^\nu}\right] = i\Box D = 0 . \qquad [15.8]$$

(This shows that one can have such an auxiliary relation only for $m=0$.)

In order to circumvent the difficulty of Eq. [15.6], we make the weaker demand (following Fermi [3])

$$\frac{\partial \Phi_\mu}{\partial x^\mu} \cdot \Psi = 0 . \qquad [15.9]$$

That is, the auxiliary condition should not be an operator identity but instead should restrict the possible states. Such a restriction has typical consequences which we will now study.

To this end, we expand in terms of a complete set of eigenfunctions,

$$\Phi_\mu(x) = \frac{1}{\sqrt{V}} \sum_{\boldsymbol{k}} \frac{1}{\sqrt{2\omega}} \sum_{\lambda=1}^{4} e_{\lambda\mu}(\boldsymbol{k}) \left[A_\lambda(\boldsymbol{k}) \exp[i(kx)] + A_\lambda^*(\boldsymbol{k}) \exp[-i(\boldsymbol{kx})]\right] , \qquad [15.10]$$

where $\Phi_4 = i\Phi_0$, $\boldsymbol{\Phi}$ and Φ_0 are Hermitian, and $\omega = +|\boldsymbol{k}|$. We want

$$(e_\lambda e_{\lambda'}) \equiv \sum_\nu e_{\lambda\nu}\, e_{\lambda'\nu} = \delta_{\lambda\lambda'} , \qquad [15.11]$$

[3] E. Fermi, *Rendiconti d. R. Acc. d. Lincei* **9**, 881 (1929); **12**, 431 (1930); *Rev. Mod. Phys.* **4**, 87 (1932); Part III.

so that e_1, e_2, e_3 and e_4 form an orthogonal basis. Furthermore,

$$\sum_\lambda e_{\lambda\mu} e_{\lambda\nu} = \delta_{\mu\nu} . \qquad [15.12]$$

Not all of the e_λ are real:

$$\begin{aligned} e_{i\mu} &= (\boldsymbol{e}_i, e_{i0}) \quad \text{is real} \qquad \text{for } i = 1, 2, 3, \\ ie_{4\mu} &= i(\boldsymbol{e}_4, e_{40}) \quad \text{is real.} \end{aligned}$$

The $e_{\lambda\mu}$ can be represented most simply as follows:

$$\left.\begin{aligned} (e_3 + ie_4)_\mu &= f(\omega) \cdot k_\mu \\ (e_1 k) = (e_2 k) &= 0 \end{aligned}\right\} . \qquad [15.13]$$

Since

$$\sum_{\mu=1}^{4} k_\mu k_\mu \equiv k^2 = 0 ,$$

then it is also true that

$$\sum_{\mu=1}^{4} (e_3 + ie_4)_\mu k_\mu = 0 .$$

However,

$$\sum_{\mu=1}^{4} (e_3 - ie_4)_\mu k_\mu \neq 0 .$$

For $\lambda = 1, 2$, the following transformation can be made:

$$\left.\begin{aligned} e'_{\lambda\mu} &= e_{\lambda\mu} + \alpha_\lambda k_\mu \qquad (\lambda = 1, 2) \\ e'_3 &= e_3 , \qquad e'_4 = e_4 \end{aligned}\right\} .$$

Because $k^2 = 0$, all equations remain correct. As an illustration, one can make the special choice

$$k_\nu = (0, 0, \omega, i\omega) ,$$

$$e_1 = (1, 0, 0, 0) ,$$

$$e_2 = (0, 1, 0, 0) ,$$

$$e_3 = (0, 0, 1, 0) ,$$

$$e_4 = (0, 0, 0, 1) .$$

The auxiliary condition [15.9] becomes

$$\left.\begin{aligned}(A_3 + iA_4)\Psi = 0\\ (A_3^* + iA_4^*)\Psi = 0\end{aligned}\right\} \qquad [15.14]$$

(cf. Eq. [15.23] below).

Hermiticity: The reality conditions for Φ_ν demand that for

$$\left.\begin{aligned}&\lambda = 1, 2, 3,\\ &\quad A_\lambda^* = \text{Hermitian conjugate of } A_\lambda = (A_\lambda)^\dagger\\ &\lambda = 4,\\ &-A_4^* = \text{Hermitian conjugate of } A_4 = (A_4)^\dagger\end{aligned}\right| . \qquad [15.15]$$

"Strong" commutation relations:

$$[A_\lambda(\boldsymbol{k}), A_{\lambda'}^*(\boldsymbol{k}')] = \delta_{\lambda\lambda'}\delta_{\boldsymbol{k},\boldsymbol{k}'} , \qquad [15.16]$$

the other commutators being zero. If we set $A_4 = B^*$, $A_4^* = -B$, then B^* is the Hermitian conjugate of B: $B^* = B^\dagger$. In addition,

$$[B, B^*] = +1 ,$$

which is the reverse of the usual situation. If, therefore, we do not wish to introduce a new metric in the Hilbert space, then we must make new interpretations:

$$\left\{\begin{aligned}&A_1,\ A_2,\ A_3,\ A_4^* \quad \text{are annihilation operators} ,\\ &A_1^*,\ A_2^*,\ A_3^*,\ A_4 \quad \text{are creation operators} ;\end{aligned}\right.$$

$$N_i = A_i^* A_i \qquad \text{for } i = 1, 2, 3,$$

$$N_4 = -A_4 A_4^* = B^* B .$$

These are the number operators, which have eigenvalues 0, 1, 2, ... Thus, for $i = 1, 2, 3$,

$$A_i \Psi(N_i) = \sqrt{N_i}\,\Psi(N_i - 1) ,$$

$$A_i^* \Psi(N_i) = \sqrt{N_i + 1}\,\Psi(N_i + 1) .$$

However, for $i=4$,

$$A_4 \Psi(N_4) = -i\sqrt{N_4+1}\,\Psi(N_4+1)\,,$$
$$A_4^* \Psi(N_4) = -i\sqrt{N_4}\,\Psi(N_4-1)\,.$$

(Here, an undetermined phase factor has been chosen arbitrarily.)

With this, the auxiliary conditions become [A-5]

$$\begin{cases} \sqrt{N_3}\,\Psi(N_3-1, N_4) + \sqrt{N_4+1}\,\Psi(N_3, N_4+1) = 0\,, \\ \sqrt{N_3+1}\,\Psi(N_3+1, N_4) + \sqrt{N_4}\,\Psi(N_3, N_4-1) = 0\,, \end{cases}$$

or

$$\begin{cases} \sqrt{N_3}\,\Psi(N_3-1, N_4-1) + \sqrt{N_4}\,\Psi(N_3, N_4) = 0\,, \\ \sqrt{N_4}\,\Psi(N_3-1, N_4-1) + \sqrt{N_3}\,\Psi(N_3, N_4) = 0\,. \end{cases}$$

The auxiliary conditions, therefore, connect states for which the difference in the N_4 quantum number is the same as the difference in the N_3 quantum number.

The general solution is

$$\Psi(N_3, N_4) = 0 \qquad \text{for } N_3 \neq N_4\,,$$
$$\Psi(N, N) = (-1)^N,$$

from which it follows that

$$\sum_N |\Psi(N, N)|^2 = \infty\,;$$

the state vector is not normalizable.

Note: This difficulty is completely independent of the vacuum; indeed we have not mentioned the vacuum at all. We will see that it is to be defined by

$$N_\lambda = A_\lambda^* A_\lambda = 0 \qquad \text{for } \lambda = 1, 2\,.$$

a. Remark on the "strong" commutation relations

The "strong" commutation relations

$$[A_\lambda(\boldsymbol{k}), A_{\lambda'}^*(\boldsymbol{k}')] = \delta_{\lambda\lambda'}\delta_{\boldsymbol{k},\boldsymbol{k}'} \qquad (\lambda, \lambda' = 1, \ldots, 4)\,, \quad [15.17]$$

are based on nongauge-invariant quantities; they are, indeed, equivalent to relations

$$[\Phi_\mu(x), \Phi_\nu(x')] = i\delta_{\mu\nu} D(x-x') , \qquad [15.18]$$

which are nongauge-invariant. That is, they are not invariant under the transformation

$$\Phi'_\mu = \Phi_\mu + \frac{\partial F}{\partial x^\mu} ,$$

with $\Box F=0$, as long as F is a q-number which has no restrictions other than $\Box F=0$.

On the other hand, we can construct "weak" commutation relations with the gauge-invariant field intensities

$$F_{\mu\nu} = \frac{\partial \Phi_\nu}{\partial x^\mu} - \frac{\partial \Phi_\mu}{\partial x^\nu} :$$

$$[F_{\mu\varrho}(x), F_{\nu\sigma}(x')] = \left(\frac{\partial^2}{\partial x^\mu \partial x^\sigma}\delta_{\varrho\nu} + \frac{\partial^2}{\partial x^\nu \partial x^\varrho}\delta_{\mu\sigma} - \frac{\partial^2}{\partial x^\mu \partial x^\nu}\delta_{\varrho\sigma} - \frac{\partial^2}{\partial x^\varrho \partial x^\sigma}\delta_{\mu\nu}\right)\cdot iD(x-x') . \qquad [15.19]$$

These are equivalent to

$$[A_\lambda(\boldsymbol{k}), A^*_{\lambda'}(\boldsymbol{k}')] = \delta_{\lambda\lambda'}\delta_{\boldsymbol{k},\boldsymbol{k}'}$$

for $\lambda=1, 2$ only. Similarly, for the vacuum expectation value we have

$$\langle\{F_{\mu\varrho}(x), F_{\nu\sigma}(x')\}\rangle_0 = -\left(\frac{\partial^2}{\partial x^\mu \partial x^\sigma}\delta_{\varrho\nu} + \frac{\partial^2}{\partial x^\nu \partial x^\varrho}\delta_{\mu\sigma} - \frac{\partial^2}{\partial x^\mu \partial x^\nu}\delta_{\varrho\sigma} - \frac{\partial^2}{\partial x^\varrho \partial x^\sigma}\delta_{\mu\nu}\right)\cdot D^{(1)}(x-x') . \qquad [15.20]$$

This is equivalent to the fact that for $\lambda=1$ and 2,

$$N_\lambda = A^*_\lambda A_\lambda , \quad \langle N_\lambda\rangle_0 = 0 , \quad \langle A_\lambda A^*_\lambda\rangle_0 = 1 .$$

Here, only the auxiliary conditions have been used.

We have seen that the "strong" commutation relations

lead to difficulties with respect to normalization of the state vector. These difficulties are inherent in such a theory, and are independent of the definition of the vacuum. They do not appear in a theory with "weak" commutation relations.

b. A Theorem of Dyson[4]

Let

$$J \equiv \int d^4x \int d^4x' K_{\mu\nu}(x, x') \Phi_\mu(x) \Phi_\nu(x') , \qquad [15.21]$$

where the gauge-invariance requirement is satisfied:

$$\frac{\partial K_{\mu\nu}}{\partial x^\mu} = \frac{\partial K_{\mu\nu}}{\partial x'^\nu} = 0 .$$

Theorem:

$$\langle J \rangle_0 = \int d^4x \int d^4x' K_{\alpha\alpha}(x, x') \cdot \tfrac{1}{2}(D^1(x - x') + i\, D(x - x')) . \qquad [15.22]$$

This can be derived on the basis of the "weak" commutation relations alone.

Proof: In momentum space, this statement is equivalent to

$$K_{\mu\nu}(k, -k) \langle A_\mu^*(\boldsymbol{k}) A_\nu(\boldsymbol{k}) \rangle_0 = 0$$

and

$$K_{\mu\nu}(k, -k) \langle A_\mu(\boldsymbol{k}) A_\nu^*(\boldsymbol{k}) \rangle_0 = K_{\alpha\alpha}(k, -k) ,$$

if only the gauge-invariance requirements

$$(E) \begin{cases} K_{\mu\nu} k_\nu = 0 , \\ k_\mu K_{\mu\nu} = 0 \end{cases}$$

are satisfied. It is useful to set

$$\frac{1}{\sqrt{2}}(e_3 + ie_4) = e_+ , \qquad \frac{1}{\sqrt{2}}(e_3 - ie_4) = e_- ,$$

with e_+ and e_- real. Then

$$(e_+ e_+) = (e_- e_-) = 0 ; \qquad (e_+ e_-) = 1 .$$

Thus, e_+ and e_- are nonorthogonal null-vectors.

[4] F. J. Dyson, *Phys. Rev.* **77**, 420 (1950).

In the following, λ takes on only the values 1 and 2. Then

$$e_{+\mu} = f(\omega) \cdot k_\mu ,$$
$$(e_\lambda e_+) = (e_\lambda e_-) = 0 ,$$
$$(e_\lambda e_{\lambda'}) = \delta_{\lambda\lambda'} .$$

Analogously, we denote

$$\frac{1}{\sqrt{2}}(A_3 \pm iA_4) = A_\pm ,$$

$$\frac{1}{\sqrt{2}}(A_3^* \pm iA_4^*) = A_\pm^* .$$

Then, $A_+^* = (A_+)^\dagger$, $A_-^* = (A_-)^\dagger$, since, indeed, $A_4^* = -(A_4)^\dagger$. The "strong" commutation relations require, in addition to the "weak" commutation relations, that

$$\begin{cases} [A_+, A_+^*] = [A_-, A_-^*] = 0 , \\ [A_+, A_-^*] = [A_-, A_+^*] = 1 . \end{cases}$$

The auxiliary conditions are

$$\begin{cases} A_+ \Psi = 0 , \\ A_+^* \Psi = 0 . \end{cases}$$

Now, any four-vector can be decomposed as follows:

$$F_\mu \sim (F_1, F_2, F_+, F_-) ,$$
$$F_\mu = \sum_{\lambda=1,2} F_\lambda e_{\lambda\mu} + F_- e_{+\mu} + F_+ e_{-\mu} ,$$

where

$$F_\lambda = (Fe_\lambda) ,$$
$$F_+ = (Fe_+) ,$$
$$F_- = (Fe_-) .$$

If $(Fk) \equiv F_\nu k_\nu = 0$, this means that

$$F_+ = 0 . \qquad [15.23]$$

That is, the term with e_- drops out.

We apply this formalism to our problem:

$$K_{\mu\nu}=\sum_{\lambda\lambda'}K_{\lambda\lambda'}e_{\lambda\mu}e_{\lambda'\nu}+K_{\lambda-}e_{\lambda\mu}e_{+\lambda'}+K_{-\lambda'}e_{+\mu}e_{\lambda'\nu}+K_{--}e_{+\mu}e_{+\nu}\,,$$

in which the gauge-invariance requirement (E) is already satisfied. Then,

$$\sum_{\mu\nu}K_{\mu\nu}(k,-k)A_\mu(\boldsymbol{k})A^*_\nu(\boldsymbol{k})$$
$$=\sum_{\lambda\lambda'}K_{\lambda\lambda'}A_\lambda A^*_{\lambda'}+K_{\lambda-}A_\lambda A^*_++K_{-\lambda}A_+A^*_\lambda+K_{--}A_+A^*_+\,.$$

Because of the auxiliary condition $A^*_+\Psi=A_+\Psi=0$, the last three terms in the vacuum expectation value drop out ($[A_+, A^*_\lambda]=0$), and there remains

$$\langle K_{\mu\nu}(k,-k)A_\mu(\boldsymbol{k})A^*_\nu(\boldsymbol{k})\rangle_0=\sum_{\lambda,\lambda'=1,2}\langle K_{\lambda\lambda'}A_\lambda A^*_{\lambda'}\rangle_0=\sum_{\lambda=1,2}K_{\lambda\lambda}\,,$$

$$\langle K_{\mu\nu}(k,-k)A^*_\mu(\boldsymbol{k})A_\nu(\boldsymbol{k})\rangle_0=\sum_{\lambda,\lambda'=1,2}\langle K_{\lambda\lambda'}A^*_\lambda A_{\lambda'}\rangle_0=0\,.$$

Furthermore, $\sum_{\lambda=1,2}K_{\lambda\lambda}=\sum_{\alpha=1}^{4}K_{\alpha\alpha}$, since $K_{33}+K_{44}=K_{+-}+K_{-+}$, and since $K_{+-}=K_{-+}=0$ because of the gauge-invariance condition (E).

Note: 1. The derivation is not completely rigorous because the terms $K_{++}\langle A_-A^*_-\rangle_0$, which we have dropped on the basis of this gauge-invariance condition (E), are actually indeterminate. This results from the fact that $A_-\Psi$ and $A^*_-\Psi$ become infinite because of the unnormalizeability, while K_{++} vanishes because of (E). Thus, these terms are actually of the form $0\times\infty$. We have, therefore, implicitly introduced a supplementary rule on how to proceed with these terms: the gauge-invariance condition (E) is to be "stronger" than the infinity of the norm of the vector.

2. Dyson's theorem is not all that is needed; we will see that, in addition,

$$J\equiv\int\mathrm{d}^4x\int\mathrm{d}^4x'\,K_{\mu\nu}(x,x')\,\varepsilon(x-x')[\Phi_\mu(x),\Phi_\mu(x')]$$

is required. This quantity is indeterminate even if $K_{\mu\nu}$ satisfies the gauge condition (E), since $K_{\mu\nu}\varepsilon(t)$ does not satisfy (E). A theorem such as

$$\langle J\rangle_0 = \int d^4x \int d^4x'\, K_{\mu\nu}(x, x')\,\varepsilon(x - x')\cdot iD(x - x')\cdot \delta_{\mu\nu}$$

can be established only with the aid of the "strong" commutation relations. The logical situation is, therefore, not satisfying, since, on the one hand, one uses the "strong" commutation relations and, on the other hand, the auxiliary conditions, which, together, lead to the nonnormalizeable states. This can be circumvented with the help of the indefinite metric.

c. *The Gupta-Bleuler Treatment of Quantum Electrodynamics*[5]

Both of these authors show that all of these difficulties can be circumvented by using the so-called "negative probability" formalism; that is, one employs an indefinite metric in Hilbert space.

Usually, the norm of a vector Ψ in Hilbert space is defined as $\sum_n \Psi_n^* \Psi_n$, where Ψ_n^* is the complex conjugate of Ψ_n. Then, if the Hamiltonian is Hermitian ($H_{nm}^* = H_{mn}$, or $H^\dagger = H$, where † denotes Hermitian conjugate), the norm remains constant in time. Expectation values are given by

$$\langle A\rangle = \frac{\sum_{mn} \Psi_n^* A_{nm} \Psi_m}{\sum_n \Psi_n^* \Psi_n}\,.$$

This is now generalized by defining, in the Hilbert space, a metric operator η, which, although Hermitian, need not be positive definite. Then, one defines the following

norm: $$\sum_{nm} \Psi_n^* \eta_{nm} \Psi_m = (\Psi^* \eta \Psi)\,,$$

expectation value: $$\langle A\rangle = \frac{(\Psi^* \eta A \Psi)}{(\Psi^* \eta \Psi)}\,.$$

Because $\eta^\dagger = \eta$, the condition for the norm to remain con-

[5] S. Gupta, *Proc. Phys. Soc. (London)* **53 A**, 681 (1950); K. Bleuler, *Helv. Phys. Acta* **23**, 567 (1950).

stant with time is

$$\eta H = H^{\dagger}\eta \, .$$

If we define an operator A^* to be the adjoint of A by

$$A^* \equiv \eta^{-1}A^{\dagger}\eta \, ,$$

and define a self-adjoint operator by

$$A^* = A \, ,$$

then the condition on the Hamiltonian for conservation of the norm is that H be self-adjoint. Furthermore, self-adjoint operators have real expectation values.

Criticism of this theory:

1. In the usual theory the norm is positive so that it can be made equal to $+1$ by multiplication by a positive number. In the theory under consideration one can, in this manner, obtain only the numbers $+1$, -1, or 0. The null states are singular, and no expectation values can be defined for them.

2. The theory is very formal, since a negative probability has no significance. One should, therefore, hardly apply such a theory to physical quantities. Nevertheless, it can be advantageous for the description of nonphysical quantities such as, for example, longitudinally polarized photons.

To this end, one proceeds as follows. The adjoint operators are defined by

$$A_\lambda^* \equiv A_\lambda^{\dagger} = A_\lambda \quad (\lambda = 1, 2, 3) \, ,$$
$$-A_4^* \equiv A_4^{\dagger} = A_4 \, ,$$

where † denotes the Hermitian conjugate. Then,

$$\Phi_\mu(x) = \frac{1}{\sqrt{V}} \sum_{\boldsymbol{k}} \frac{1}{\sqrt{2\omega}} \sum_\lambda e_{\mu\lambda}(\boldsymbol{k}) \cdot [A_\lambda(\boldsymbol{k}) \exp[i(kx)] + A_\lambda^{\dagger}(\boldsymbol{k}) \exp[-i(kx)]] \, .$$

Thus, $\Phi_\mu(x)$ is not Hermitian, but is self-adjoint instead.

The commutation relation

$$[A_\lambda(\boldsymbol{k}), A^\dagger_{\lambda'}(\boldsymbol{k}')] = \delta_{\lambda\lambda'}\,\delta_{\boldsymbol{k},\boldsymbol{k}'}$$

remains valid. For $\lambda = 1, 2, 3, 4$, then,

$$A_\lambda = \text{annihilation operator},$$

$$A^\dagger_\lambda = \text{creation operator},$$

$$N_\lambda = A^\dagger_\lambda A_\lambda = \text{number operator}.$$

The quantity [A-5]

$$\sum_{(N_\lambda)} \Psi^*(N_\lambda)\Psi(N_\lambda)(-1)^{N_4}$$

remains constant in time. We can then demand that

$$\begin{cases} \eta A^\dagger_\lambda = A^\dagger_\lambda \eta & (\lambda = 1, 2, 3), \\ \eta A^\dagger_4 = -A^\dagger_4 \eta . \end{cases}$$

From the matrix representation

$$(N_4|\eta|N_4)(N_4|A_4|N_4 - 1) = -(N_4|A_4|N_4 - 1)(N_4 - 1|\eta|N_4 - 1),$$

it can be seen that the solution is

$$(N_4|\eta|N_4') = \delta_{N_4 N_4'}(-1)^{N_4}.$$

We must weaken the auxiliary condition, since

$$(A^\dagger_3 + iA^\dagger_4)\Psi = 0$$

cannot be satisfied. For this reason we demand the auxiliary condition only for the positive frequencies:

$$\left(\frac{\partial \Phi_\mu}{\partial x^\mu}\right)^+ \Psi = 0 .$$

This restriction is not dangerous since, for the expectation value,

$$\left\langle \frac{\partial \Phi_\mu}{\partial x^\mu} \right\rangle = 0$$

remains satisfied, as one can easily convince oneself.

If we go over to $A_\pm$ as before, then it is to be noted that now

$$A_\pm^\dagger = \frac{1}{\sqrt{2}}(A_3^\dagger \mp iA_4^\dagger)\,.$$

Thus, for $\boldsymbol{k} = \boldsymbol{k}'$,

$$[A_+, A_+^\dagger] = [A_-, A_-^\dagger] = 1\,,$$
$$[A_+, A_-^\dagger] = [A_-, A_+^\dagger] = 0\,,$$

which is the opposite of what we had before. It is then easily seen that for every state the expectation values become

$$\langle A_-^\dagger A_+\rangle = \langle A_+^\dagger A_-\rangle = \langle A_+ A_-^\dagger\rangle = \langle A_- A_+^\dagger\rangle = \langle A_+^\dagger A_+\rangle = 0\,,$$
$$\langle A_+ A_+^\dagger\rangle = 1\,,$$

and

$$\langle A_- A_-^\dagger\rangle = 1 - \langle A_-^\dagger A_-\rangle\,,$$

which is indeterminate but finite.

Thus, the problem with Dyson's theorem is settled, since terms of the form $0\times\infty$ no longer appear, and since for the physical quantities only the determinate expectation value is essential.

In the following, the expectation value of Dyson's P-symbol (chronological, or time-ordered product) will be essential.

Definition:

$$P\big(A(x)B(x')\big) = \begin{Bmatrix} A(x)B(x') & (t>t') \\ B(x')A(x) & (t<t') \end{Bmatrix}, \qquad [15.24]$$

and analogously for more factors. Verbally, the P-product describes the chronological ordering of the operators according to decreasing times.

For two factors, specifically,

$$P\big(A(x)B(x')\big) = \tfrac{1}{2}\{A(x), B(x')\} + \tfrac{1}{2}\varepsilon(x-x')[A(x), B(x')]\,. \qquad [15.25]$$

In the theory of neutral scalar fields we have, for example,

$$\langle\{\psi(x), \psi(x')\}\rangle_0 = \Delta^1(x-x') ,$$
$$[\psi(x), \psi(x')] = i\Delta(x-x') .$$

Thus,

$$\langle P(\psi(x)\,\psi(x'))\rangle_0 = \tfrac{1}{2}\Delta^1(x-x') - i\bar{\Delta}(x-x')$$
$$= \tfrac{1}{2}\Delta^C(x-x') .$$

(Note that Δ^C appears here.)

It is analogous in electrodynamics. In order to avoid statements about nongauge-invariant quantities, one considers only the term that appears in applications:

$$\langle J\rangle_0 = \int \mathrm{d}^4x \int \mathrm{d}^4x'\, K_{\mu\nu}(x, x') \langle P(\Phi_\mu(x)\ \Phi_\nu(x'))\rangle_0 , \qquad [15.26]$$

with

$$\frac{\partial K_{\mu\nu}}{\partial x^\mu} = \frac{\partial K_{\mu\nu}}{\partial x'^\nu} = 0 .$$

With regard to this, Dyson's theorem states:

$$\langle J\rangle_0 = \int \mathrm{d}^4x \int \mathrm{d}^4x'\, K_{\alpha\alpha}(x, x') \cdot \tfrac{1}{2} D^C(x-x') . \qquad [15.27]$$

The indefinite metric is, for us, only a means with which to prove this theorem; we do not need it further.

Note: 1. Clearly, only vacuum expectation values of gauge-invariant quantities are well defined, that is, expressions of the type

$$\langle P\,(F_{\mu\nu}(x)\, F_{\varrho\sigma}(x'))\rangle_0 .$$

2. In this Dyson P-product, a special role is given to simultaneity, that is, to surfaces $t =$ constant. (Instead of such planes one can also, more generally, employ curved time-like surfaces which, here and there, may be practical and which, physically, are neither better nor worse than the planes. See Section 21.) The basis of this special role for simultaneity lies in the canonical formalism.

16. QUANTUM ELECTRODYNAMICS IN CANONICAL NOTATION

We derive electrodynamics from a variational principle:

$$\delta\int L\,\mathrm{d}t = 0\,,$$

where

$$L = \int \mathscr{L}\,\mathrm{d}^3x\,,$$

$$\left.\begin{aligned}\mathscr{L} &= -\frac{1}{2}\frac{\partial\Phi_\mu}{\partial x^\nu}\frac{\partial\Phi_\mu}{\partial x^\nu}\\ &= \frac{1}{2}\left\{\left[\left(\frac{\partial\mathbf{\Phi}}{\partial t}\right)^2 - \sum_k\left(\frac{\partial\mathbf{\Phi}}{\partial x^k}\right)^2\right] - \left[\left(\frac{\partial\Phi_0}{\partial t}\right)^2 - \sum_k\left(\frac{\partial\Phi_0}{\partial x^k}\right)^2\right]\right\}\end{aligned}\right\}. \qquad [16.1]$$

The canonically conjugate momenta are

$$P(x) = \frac{\delta L}{\delta\dot{Q}} = \frac{\partial\mathscr{L}}{\partial\dot{Q}}\,,$$

$$\left.\begin{aligned}\boldsymbol{P}(x) &= \frac{\partial\mathbf{\Phi}}{\partial t}\\ P_0(x) &= -\frac{\partial\Phi_0}{\partial t}\end{aligned}\right\}. \qquad [16.2]$$

The canonical commutation relations are

$$\left.\begin{aligned}i[P_i(\boldsymbol{x}, t), \Phi_k(\boldsymbol{x}', t)] &= \delta_{ik}\delta^3(\boldsymbol{x}-\boldsymbol{x}')\\ i[P_0(\boldsymbol{x}, t), \Phi_0(\boldsymbol{x}', t)] &= +\delta^3(\boldsymbol{x}-\boldsymbol{x}')\end{aligned}\right\},$$

all others being zero.

These relations are valid only at equal times; the canonical formalism, from its nature, describes only what occurs at equal times.

With

$$P_4 \equiv \frac{\partial\Phi_4}{\partial t} = -iP_0\,, \qquad \Phi_4 = +i\Phi_0\,,$$

we have

$$[P_4, \Phi_4] = +[P_0, \Phi_0]\,,$$

from which

$$i[P_\mu(\boldsymbol{x}, t), \Phi_\nu(\boldsymbol{x}', t)] = i\left[\frac{\partial \Phi_\mu(\boldsymbol{x}, t)}{\partial t}, \Phi_\nu(\boldsymbol{x}', t)\right]$$
$$= \delta_{\mu\nu}\delta^3(\boldsymbol{x} - \boldsymbol{x}') \,. \qquad [16.3]$$

These commutation relations are, naturally, contained in the previous, more general relations

$$[\Phi_\mu(x), \Phi_\nu(x')] = i\delta_{\mu\nu}D(x - x') \,,$$

as can easily be checked by a calculation analogous to that in Section 14. The Hamiltonian is

$$H = \sum \dot{Q}\,\frac{\delta L}{\delta \dot{Q}} - L \equiv \int \mathscr{H}\mathrm{d}^3x \,.$$

Hence,

$$\mathscr{H} = \frac{1}{2}\left\{\left[\left(\frac{\partial \mathbf{\Phi}}{\partial t}\right)^2 + \sum_k \left(\frac{\partial \mathbf{\Phi}}{\partial x^k}\right)^2\right] - \left[\left(\frac{\partial \Phi_0}{\partial t}\right)^2 + \sum_k \left(\frac{\partial \Phi_0}{\partial x^4}\right)^2\right]\right\} \,. \qquad [16.4]$$

We still need an auxiliary condition to ensure that the expectation values satisfy Maxwell's equations. This condition does not follow from the canonical formulas. We postulate

$$\frac{\partial \Phi_\mu}{\partial x^\mu}\Psi = \left(\frac{\partial \Phi_0}{\partial t} + \operatorname{div}\mathbf{\Phi}\right)\Psi = 0 \,. \qquad [16.5]$$

The canonical equations

$$\frac{\partial \Phi_\mu(x)}{\partial t} = i[H, \Phi_\mu(x)]$$

and

$$\frac{\partial^2 \Phi_\mu(x)}{\partial t^2} = i\left[H, \frac{\partial \Phi_\mu(x)}{\partial t}\right],$$

together with the auxiliary condition, when applied to Ψ, yield the Maxwell equations.

Compatibility of the auxiliary condition: It must be

true that

$$\left[H, \frac{\partial \Phi_\mu}{\partial x^\mu}\right] \Psi = 0 \qquad [16.6]$$

(necessary condition). With the field intensity

$$\boldsymbol{E} \equiv -\nabla \Phi_0 - \frac{\partial \boldsymbol{\Phi}}{\partial t},$$

this means that div $\boldsymbol{E} \cdot \Psi = 0$. These two conditions,

$$\frac{\partial \Phi_\mu}{\partial x^\mu} \cdot \Psi = 0 \quad \text{and} \quad \left[H, \frac{\partial \Phi_\mu}{\partial x^\mu}\right] \Psi = 0,$$

are also sufficient, since if they are satisfied at one time, they are also satisfied for all times.

17. VARIOUS REPRESENTATIONS

The following is valid for ordinary quantum mechanics as well as for the theory of quantized fields.

Definition: Here $\dot{F}$ is the operator for which

$$\langle \dot{F} \rangle = \frac{\mathrm{d}}{\mathrm{d}t} \langle F \rangle . \qquad [17.1]$$

It is only a matter of convention how much of the time dependence in this is contained in the operator and how much in the Ψ function, because physical quantities are the expectation values,

$$\langle F \rangle = \sum_{mn} (\Psi_n^* F_{nm} \Psi_m) . \qquad [17.2]$$

For systems without interaction, there are essentially only two representations to be considered.

1. The Heisenberg representation:

$$\begin{cases} \dfrac{\partial \Psi_H}{\partial t} = 0 , \\[2ex] \dot{F} = \dfrac{\partial F}{\partial t} . \end{cases}$$

2. The Schrödinger representation:

$$\begin{cases} \dfrac{\partial \Psi_S}{\partial t} = -iH\Psi_S, \\ \dfrac{\partial F}{\partial t} = 0. \end{cases}$$

(In Sections 14, 15, and 16, the Heisenberg representation is always referred to. However, all of the results are also valid in the Schrödinger representation if $\partial F/\partial t$ is everywhere replaced by $\dot{F}$.)

A transformation U connects the two representations:

$$\Psi_H = U\Psi_S, \qquad [17.3]$$

$$UU^\dagger = U^\dagger U = 1,$$

where

$$\frac{\partial U}{\partial t} = +iH_S U,$$

and $H_S = H$ in the Schrödinger representation. Thus,

$$U = \exp[iH_S t], \qquad [17.4]$$

$$\langle F \rangle = (\Psi_H, F_H \Psi_H) = (\Psi_S, F_S \Psi_S).$$

Here, (Ψ_1, Ψ_2) denotes the scalar product in Hilbert space. Also,

$$F_H = UF_S U^{-1}. \qquad [17.5]$$

18. THEORY OF POSITRONS (SPIN-½ PARTICLES) [A-4]

The Dirac equation is

$$\left(\gamma^\nu \frac{\partial}{\partial x^\nu} + m\right)\psi = 0.$$

Let

$$S \equiv \left(\gamma^\nu \frac{\partial}{\partial x^\nu} - m\right)\Delta\,, \quad \text{so that} \quad \left(\gamma^\nu \frac{\partial}{\partial x^\nu} + m\right) S = 0\,,$$

$$\bar{S} \equiv \left(\gamma^\nu \frac{\partial}{\partial x^\nu} - m\right)\bar{\Delta}\,, \quad \text{so that} \quad \left(\gamma^\nu \frac{\partial}{\partial x^\nu} + m\right) \bar{S} = -\,\delta^4(x)\,,$$

$$S^1 \equiv \left(\gamma^\nu \frac{\partial}{\partial x^\nu} - m\right)\Delta^1, \quad \text{so that} \quad \left(\gamma^\nu \frac{\partial}{\partial x^\nu} + m\right) S^1 = 0\,.$$

Then (cf. Eq. [3.26])

$$\{\psi_\alpha(x),\, \bar{\psi}_\beta(x')\} = -\,iS_{\alpha\beta}(x - x')\,. \qquad [18.1]$$

In particular,

$$\{\psi_\alpha(\boldsymbol{x}, t),\, \bar{\psi}_\beta(\boldsymbol{x}', t)\} = -\,iS_{\alpha\beta}(\boldsymbol{x} - \boldsymbol{x}', 0) = +\,\gamma^4_{\alpha\beta}\delta^3(\boldsymbol{x} - \boldsymbol{x}')\,.$$

Furthermore, with

$$\bar{\psi} \equiv \psi^* \gamma^4\,,$$

$$\bar{\psi}\left(\gamma^\nu \frac{\overleftarrow{\partial}}{\partial x^\nu} - m\right) = 0\,,$$

we have

$$\{\psi_\alpha(\boldsymbol{x}, t),\, \psi^*_\beta(\boldsymbol{x}', t)\} = \delta_{\alpha\beta}\delta^3(\boldsymbol{x} - \boldsymbol{x}')\,. \qquad [18.2]$$

This, together with the Dirac equation, is equivalent to the general commutation relation of Eq. [18.1].

Canonical formalism

The momenta π are unnatural here, since at a given time the π's and ψ's cannot be given independently of one another. Indeed, $\dot{\psi}$ follows from ψ via the Dirac equation. On the other hand, the ψ's satisfy no auxiliary condition; they can be freely specified beforehand. The equations

$$\frac{\partial \psi}{\partial t} = i[H, \psi]\,, \qquad \frac{\partial \psi^*}{\partial t} = i[H, \psi^*]\,,$$

can be easily satisfied by

$$H = \int \mathscr{H}\, d^3x \, .$$

In order to specify $\mathscr{H}$, we go over to the $\boldsymbol{\alpha}$ matrices:

$$\alpha_k = i\gamma^4\gamma^k \qquad (k = 1, 2, 3) \, ,$$

$$\beta = \gamma^4 \, .$$

The Dirac equation is then

$$\frac{\partial\psi}{\partial t} + \boldsymbol{\alpha}\cdot\frac{\partial\psi}{\partial \boldsymbol{x}} + i\beta m\psi = 0 \, .$$

For $\mathscr{H}$, we choose

$$\mathscr{H} = -\frac{i}{2}\left\{\psi^*\left(\boldsymbol{\alpha}\cdot\frac{\partial\psi}{\partial\boldsymbol{x}} + i\beta m\psi\right) - \left(\frac{\partial\psi^*}{\partial\boldsymbol{x}}\cdot\boldsymbol{\alpha} - i\psi^*\beta m\right)\psi\right\} .$$

The sequence of factors is just right. This hypothesis satisfies our requirement, as can easily be shown.

Note: 1. The commutation relations [18.2] contain the anticommutator; the equations of motion, however, contain the commutator. This turns out correctly because $\mathscr{H}$ is bilinear in ψ.

2. Formally, one can start from a Lagrangian function:

$$-\mathscr{L} = \frac{1}{2}\left(\bar\psi\gamma^\nu\frac{\partial\psi}{\partial x^\nu} - \frac{\partial\bar\psi}{\partial x^\nu}\gamma^\nu\psi\right) + m\bar\psi\psi$$

$$= -\frac{i}{2}\left\{\psi^*\left(\frac{\partial\psi}{\partial t} + \boldsymbol{\alpha}\cdot\frac{\partial\psi}{\partial\boldsymbol{x}} + i\beta m\psi\right) - \left(\frac{\partial\psi^*}{\partial t} + \frac{\partial\psi^*}{\partial\boldsymbol{x}}\cdot\boldsymbol{\alpha} - im\psi^*\beta\right)\psi\right\} .$$

The Dirac equation then follows from $\delta\int\mathscr{L}\, d^3x = 0$. However, $\mathscr{L} = 0$ (along an extremal path) also follows, and this is the degeneracy which appears here.

Chapter 4. Interacting Fields: Interaction Representation and S-Matrix

19. ELECTRONS INTERACTING WITH THE ELECTROMAGNETIC FIELD

Here, we must make the substitutions

$$\frac{\partial\psi}{\partial x^\nu} \to \frac{\partial\psi}{\partial x^\nu} - ie\Phi_\nu\psi\,,$$

$$\frac{\partial\psi^*}{\partial x^\nu} \to \frac{\partial\psi^*}{\partial x^\nu} + ie\Phi_\nu\psi^*\,.$$

Then, the field equations read

$$\left.\begin{aligned} \frac{\partial\psi}{\partial t} &= i[H + H_{\text{int}}, \psi] \\ \frac{\partial\psi^*}{\partial t} &= i[H + H_{\text{int}}, \psi^*] \end{aligned}\right\}, \qquad [19.1]$$

where H is the previous H, and where

$$\left.\begin{aligned} H_{\text{int}} &= \int \mathscr{H}_{\text{int}}\, \mathrm{d}^3x \\ \mathscr{H}_{\text{int}} &= -\, j^\nu\,\Phi_\nu = j^0\,\Phi_0 - \boldsymbol{j}\cdot\boldsymbol{\Phi} \end{aligned}\right\}, \qquad [19.2]$$

in which

$$j^\nu = ie\bar{\psi}\gamma^\nu\psi\,,$$

$$j^0 = e\psi^*\psi\,, \qquad \boldsymbol{j} = e\psi^*\boldsymbol{\alpha}\psi\,.$$

Thus,

$$\left.\begin{aligned} &\frac{\partial\psi}{\partial t} + ie\Phi_0\psi + \boldsymbol{\alpha}\cdot\left(\frac{\partial\psi}{\partial\boldsymbol{x}} - ie\boldsymbol{\Phi}\psi\right) + i\beta m\psi = 0 \\ &\frac{\partial\psi^*}{\partial t} - ie\Phi_0\psi + \left(\frac{\partial\psi^*}{\partial\boldsymbol{x}} + ie\boldsymbol{\Phi}\psi^*\right)\cdot\boldsymbol{\alpha} - im\psi^*\beta = 0 \end{aligned}\right\}. \qquad [19.3]$$

Note: 1. Formally speaking, this is based on the Heisenberg representation. Nevertheless, it is the same in the Schrödinger representation if $\partial\psi/\partial t$ is replaced by $\dot{\psi}$ (see Section 17).

2. A special feature here is that $\mathscr{H}_{\text{int}}$ is Lorentz invariant and contains no derivatives of the participating fields.

3. Commutation relations: nothing changes in the canonical commutation relations (at equal times). This is the advantage of the canonical formulation. However, they can no longer be extended to different times.

4. If one considers electrons interacting with a quantized electromagnetic field, then one must write for the Hamiltonian

$$H = H_{\text{em}} + H_{\text{Dirac}} + H_{\text{int}} ,$$

where everything else remains unchanged.

20. CHARGED PARTICLES WITH ZERO SPIN[1]

We have

$$\left.\begin{aligned} \mathscr{L}_m^0 &= -\frac{\partial\psi^*}{\partial x^\mu}\frac{\partial\psi}{\partial x^\mu} - m^2\psi^*\psi \\ &= \frac{\partial\psi^*}{\partial t}\frac{\partial\psi}{\partial t} - \frac{\partial\psi^*}{\partial \boldsymbol{x}}\cdot\frac{\partial\psi}{\partial \boldsymbol{x}} - m^2\psi^*\psi \end{aligned}\right\} . \qquad [20.1]$$

From

$$\left.\begin{aligned} \pi &\equiv \frac{\delta L}{\delta(\partial\psi/\partial t)} = \frac{\partial\mathscr{L}}{\partial(\partial\psi/\partial t)} = \frac{\partial\psi^*}{\partial t} \\ \pi^* &= \frac{\partial\psi}{\partial t} \end{aligned}\right\} , \qquad [20.2]$$

there results

$$\left.\begin{aligned} & i[\pi(\boldsymbol{x}, t), \psi(\boldsymbol{x}', t)] \\ &\qquad = i\left[\frac{\partial\psi^*}{\partial t}(\boldsymbol{x}, t), \psi(\boldsymbol{x}', t)\right] = \delta^3(\boldsymbol{x} - \boldsymbol{x}') \\ & i[\pi^*(\boldsymbol{x}, t), \psi^*(\boldsymbol{x}', t)] \\ &\qquad = i\left[\frac{\partial\psi}{\partial t}(\boldsymbol{x}, t), \psi^*(\boldsymbol{x}', t)\right] = \delta^3(\boldsymbol{x} - \boldsymbol{x}') \end{aligned}\right\} , \qquad [20.3]$$

[1] See Section 9 for a formulation using Δ functions.

and

$$\mathscr{H}^0_m = \pi\pi^* + \frac{\partial\psi^*}{\partial\boldsymbol{x}}\cdot\frac{\partial\psi}{\partial\boldsymbol{x}} + m^2\psi^*\psi\,. \qquad [20.4]$$

The field equations become

$$\left(\frac{\partial}{\partial x^\nu}\frac{\partial}{\partial x^\nu} - m^2\right)\psi = 0\,. \qquad [20.5]$$

Interaction with the electromagnetic field

The usual substitutions,

$$\frac{\partial\psi}{\partial x^\nu} \to \frac{\partial\psi}{\partial x^\nu} - ie\Phi_\nu\psi\,,$$

$$\frac{\partial\psi^*}{\partial x^\nu} \to \frac{\partial\psi^*}{\partial x^\nu} + ie\Phi_\nu\psi^*\,,$$

are made in all formulas. The field equations become

$$\left.\begin{aligned}&\left(\frac{\partial}{\partial x^\nu} - ie\Phi_\nu\right)\left(\frac{\partial}{\partial x^\nu} - ie\Phi_\nu\right)\psi - m^2\psi = 0\\ &\left(\frac{\partial}{\partial x^\nu} + ie\Phi_\nu\right)\left(\frac{\partial}{\partial x^\nu} + ie\Phi_\nu\right)\psi^* - m^2\psi^* = 0\end{aligned}\right\}. \qquad [20.6]$$

The corresponding variational principle is

$$\delta\int\mathscr{L}\,\mathrm{d}^4x = 0\,,$$

where

$$\mathscr{L} = \mathscr{L}^0 + \mathscr{L}^{\text{int}} = -\left(\frac{\partial\psi^*}{\partial x^\mu} + ie\Phi_\mu\psi^*\right)\cdot\left(\frac{\partial\psi}{\partial x^\mu} - ie\Phi_\mu\psi\right) - m^2\psi^*\psi\,. \qquad [20.7]$$

Separating space and time,

$$\mathscr{L} = +\left(\frac{\partial\psi^*}{\partial t} - ie\Phi_0\psi^*\right)\left(\frac{\partial\psi}{\partial t} + ie\Phi_0\psi\right) - \left(\frac{\partial\psi^*}{\partial\boldsymbol{x}} + ie\boldsymbol{\Phi}\psi^*\right)\cdot\left(\frac{\partial\psi}{\partial\boldsymbol{x}} - ie\boldsymbol{\Phi}\psi\right) - m^2\psi^*\psi\,. \qquad [20.8]$$

In this case, the Hamiltonian formalism functions quite normally. With

$$L=\int \mathscr{L} \mathrm{d}^3x \,,$$

we have

$$\pi=\frac{\delta L}{\delta(\partial\psi/\partial t)}=\frac{\partial\mathscr{L}}{\partial(\partial\psi/\partial t)}=\frac{\partial\psi^*}{\partial t}-ie\Phi_0\psi^* .$$

That is, as compared to the case without interaction, here π contains an additional term:

$$\left.\begin{aligned}\pi\; &=\frac{\partial\psi^*}{\partial t}-ie\Phi_0\psi^*\\ \pi^* &=\frac{\partial\psi}{\partial t}+ie\Phi_0\psi\end{aligned}\right\} . \qquad [20.9]$$

The canonical commutation relations are

$$\left.\begin{aligned} i[\pi(\boldsymbol{x},t),\psi(\boldsymbol{x}',t)]=i[\pi^*(\boldsymbol{x},t),\psi^*(\boldsymbol{x}',t)]=\delta^3(\boldsymbol{x}-\boldsymbol{x}')\\ \text{all others}=0\end{aligned}\right\} . \qquad [20.10]$$

Note: 1. π and not $\partial\psi/\partial t$ has simple commutation relations. 2. The relativistic invariance is anything but obvious.

The Hamiltonian is

$$\mathscr{H}=\pi\frac{\partial\psi}{\partial t}+\pi^*\frac{\partial\psi^*}{\partial t}-\mathscr{L} .$$

This is to be expressed in terms of π and ψ, rather than $\partial\psi/\partial t$ and ψ:

$$\begin{aligned}\mathscr{H}&=\pi\pi^*+m^2\psi^*\psi+ie\Phi_0(\pi^*\psi^*-\pi\psi)\\ &\qquad+\left(\frac{\partial\psi^*}{\partial\boldsymbol{x}}+ie\boldsymbol{\Phi}\psi^*\right)\cdot\left(\frac{\partial\psi}{\partial\boldsymbol{x}}-ie\boldsymbol{\Phi}\psi\right) .\end{aligned} \qquad [20.11]$$

We write

$$\mathscr{H}=\mathscr{H}_0+\mathscr{H}_{\text{int}} .$$

Here $\mathscr{H}_0$ is identical with that written down previously:

$$\mathscr{H}_0 = \pi\pi^* + m^2\psi^*\psi + \frac{\partial\psi^*}{\partial \boldsymbol{x}}\cdot\frac{\partial\psi}{\partial \boldsymbol{x}},$$

$$\mathscr{H}_{\text{int}} = ie\Phi_0(\pi^*\psi^* - \pi\psi) + ie\boldsymbol{\Phi}\cdot\left(\psi^*\frac{\partial\psi}{\partial \boldsymbol{x}} - \frac{\partial\psi^*}{\partial \boldsymbol{x}}\psi\right) + e^2\boldsymbol{\Phi}^2\psi^*\psi\,.$$

The current density is

$$j^\mu \equiv \frac{\partial\mathscr{L}_{\text{mat}}}{\partial\Phi_\mu},$$

because of Maxwell's equation. Thus,

$$j^\mu = ie\left(\frac{\partial\psi^*}{\partial x^\mu}\psi - \psi^*\frac{\partial\psi}{\partial x^\mu}\right) - 2e^2\Phi_\mu\psi^*\psi\,. \qquad [20.12]$$

The equation of continuity is satisfied:

$$\frac{\partial j^\mu}{\partial x^\mu} = 0\,, \qquad [20.13]$$

$$\boldsymbol{j} = ie\left(\frac{\partial\psi^*}{\partial \boldsymbol{x}}\psi - \psi^*\frac{\partial\psi}{\partial \boldsymbol{x}}\right) - 2e^2\boldsymbol{\Phi}\psi^*\psi\,.$$

The charge density is

$$\begin{aligned} j^0 &= -ie\left(\frac{\partial\psi^*}{\partial t}\psi - \psi^*\frac{\partial\psi}{\partial t}\right) - 2e^2\Phi_0\psi^*\psi \\ &= +ie(\pi^*\psi^* - \pi\psi)\,. \end{aligned}$$

Note: 1. To begin with, this is true for an external electromagnetic field. If one adds the radiation Hamiltonian and keeps the commutation relations between canonically conjugate quantities unchanged, then it is also true for the quantized radiation. In addition, in the Schrödinger and Heisenberg representations considered here, the auxiliary condition need not be changed.

2. With respect to the auxiliary condition, it is to be noted that

$$\begin{aligned} \left[\left(\frac{\partial\Phi_\mu}{\partial x^\mu}\right)_{\boldsymbol{x},t}, \mathscr{H}_{\text{int}}(\boldsymbol{x}', t)\right] &= \left[\frac{\partial\Phi_0(\boldsymbol{x}, t)}{\partial t}, \Phi_0(\boldsymbol{x}', t)\right] j_0(\boldsymbol{x}', t) \\ &= ij_0(\boldsymbol{x}, t)\,\delta^3(\boldsymbol{x} - \boldsymbol{x}')\,. \end{aligned} \qquad [20.14]$$

This equation is also valid for spin-$\frac{1}{2}$:

$$\mathscr{H}_{\text{int}} = -j^{\nu}\Phi_{\nu} = j^{0}\Phi_{0} - \boldsymbol{j}\cdot\boldsymbol{\Phi} .$$

What is always essential is the term with Φ_0, and this appears as $+j^0\Phi_0$ both times.

21. THE INTERACTION REPRESENTATION

This representation, introduced by Tomonaga and Schwinger,[2] lies, so to speak, between the Schrödinger and the Heisenberg representations.

In Section 17 we have seen that how much time dependence is ascribed to the state vectors Ψ and how much to the operators F is a matter of convention, as long as the expectation values $\langle F\rangle = (\Psi^* F\Psi)$ have the proper time dependence:

$$\langle\dot{F}\rangle = \frac{\mathrm{d}}{\mathrm{d}t}\langle F\rangle .$$

In the interaction representation, the operators have the time dependence of the force-free case, and they satisfy the force-free field equations. The Ψ, then, contains a time dependence which makes the time dependence of the expectation values correct. The commutation relations of the operators are also the same as the force-free ones.

Dirac, Fock, and Podolski[3] had already done something similar, but they treated matter and radiation on different terms.

We demand, then, for an operator O_i in the interaction representation, that

$$\frac{\partial O_i}{\partial t} = i[H_0, O_i] . \qquad [21.1]$$

[2] S. Tomonaga, *Progr. Theor. Phys.* **1**, 27, 109 (1946); J. Schwinger, *Phys. Rev.* **74**, 1439 (1948); **75**, 651 (1949).

[3] P. A. M. Dirac, V. A. Fock, and B. Podolsky, *Phys. Zeitschrift der Sowjetunion* **2**, Heft 6 (1936).

Furthermore, we define

$$\dot{O}_i \equiv i[H_0 + H_{\text{int}}, O_i]\,. \tag{21.2}$$

Then

$$\begin{aligned}\frac{\mathrm{d}}{\mathrm{d}t}\langle O_i\rangle &= \frac{\mathrm{d}}{\mathrm{d}t}(\Psi_i, O_i\Psi_i)\\ &= i\big(\Psi_i, ((H_0 + H_{\text{int}})O_i - O_i(H_0 + H_{\text{int}}))\Psi_i\big)\\ &= \left(\Psi_i, \frac{\partial O_i}{\partial t}\Psi_i\right) + i\big(\Psi_i, (H_{\text{int}}O_i - O_i H_{\text{int}})\Psi_i\big)\,,\end{aligned}$$

which implies

$$\frac{\partial \Psi_i}{\partial t} = -\,iH_{\text{int}}\Psi_i\,. \tag{21.3}$$

The interaction representation Ψ_i is obtained from Schrödinger's Ψ_S and Heisenberg's Ψ_H by means of a unitary transformation:

$$\Psi_i = \exp[iH_0 t]\Psi_S = U\Psi_H\,, \qquad UU^\dagger = U^\dagger U = 1\,, \tag{21.4}$$

where

$$\frac{\partial U}{\partial t} = -\,iH_{\text{int}}\,U, \tag{21.5}$$

and

$$O_i = \exp[iH_0 t]O_S \exp[-iH_0 t] = UO_H U^{-1}. \tag{21.6}$$

This representation singles out (as does the Schrödinger representation) a manifold of surfaces, $t = \text{const}$. Instead of the surfaces $t = \text{const}$, one can consider more general (curved) surfaces σ which, as their only restriction, have everywhere a time-like normal direction. Instead of $\partial\psi/\partial t$, one considers the following quantity:

$$\frac{\Psi_i(\sigma') - \Psi_i(\sigma)}{\Omega}\,,$$

where Ω denotes the finite volume enclosed between σ and σ'. Then, let

$$\frac{\delta\Psi(\sigma)}{\delta\Omega(x)} \equiv \lim_{\Omega\to x}\frac{\Psi_i(\sigma') - \Psi_i(\sigma)}{\Omega}\,, \tag{21.7}$$

in which $\Omega(x)$ in the limit shrinks to the point x. Then, it is seen that the generalization of

$$\frac{\partial \Psi_i}{\partial t} = - iH_{\text{int}} \Psi_i$$

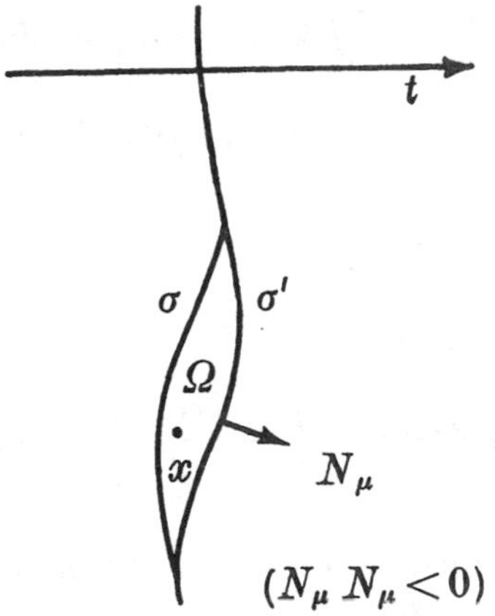

Figure 21.1

with

$$H_{\text{int}} = \int \mathscr{H}_{\text{int}} d^3x = \int \mathscr{H}_{\text{int}} d\sigma$$

becomes

$$\frac{\delta \Psi(\sigma)}{\delta \Omega(x)} = - i\mathscr{H}_{\text{int}}(x) \Psi(\sigma) .$$

Note: 1. The quantity $\mathscr{H}_{\text{int}}(x)$ can (for example, for the case of zero spin, see below), in addition to x, still depend upon the direction of N_μ, the normal to the surface σ at the point x: $\mathscr{H}_{\text{int}}(x, N_\mu)$.

2. These curved surfaces are not nice, but they are no more unphysical than the surfaces $t = \text{const}$. What is unphysical are the instantaneous quantities.

For the unitary transformation U which leads from the Heisenberg to this generalized interaction representation, it is then true that

$$\frac{\delta U}{\delta \Omega(x)} = -i\mathscr{H}_{\text{int}}\,U\,. \tag{21.8}$$

For spin-$\frac{1}{2}$, we have

$$\mathscr{H}_{\text{int}} = -j_\nu \Phi_\nu = -ie\bar{\psi}\gamma^\nu\psi\Phi_\nu\,. \tag{21.9}$$

For spin-0, the situation is less banal (see above, Note 1):

$$\left.\begin{aligned} \mathscr{H}_0 &= \pi\pi^* + \frac{\partial\psi^*}{\partial \boldsymbol{x}}\cdot\frac{\partial\psi}{\partial \boldsymbol{x}} + m^2\psi^*\psi \\ \mathscr{H}_{\text{int}} &= ie\Phi_0(\pi^*\psi^* - \pi\psi) \\ &\qquad + ie\boldsymbol{\Phi}\cdot\left(\psi^*\frac{\partial\psi}{\partial \boldsymbol{x}} - \frac{\partial\psi^*}{\partial \boldsymbol{x}}\psi\right) + e^2\boldsymbol{\Phi}^2\psi^*\psi \end{aligned}\right\}. \tag{21.10}$$

This is true in the Heisenberg representation. Naturally, the same thing results in the interaction representation; every operator is simply taken in the interaction representation. Here, however, the force-free field equations and commutation relations obtain. Thus,

$$(\pi)_i = \left(\frac{\partial\psi^*}{\partial t}\right)_i \qquad (\neq \dot{\psi}^*\,!)\,,$$

$$(\pi^*)_i = \left(\frac{\partial\psi}{\partial t}\right)_i.$$

With this, in the interaction representation, there results

$$\left.\begin{aligned} \mathscr{H}_{\text{int}} &= ie\Phi_\nu\left(\psi^*\frac{\partial\psi}{\partial x^\nu} - \frac{\partial\psi^*}{\partial x^\nu}\psi\right) + e^2\sum_1^3 \Phi_k^2\psi^*\psi \\ &= ie\Phi_\nu\left(\psi^*\frac{\partial\psi}{\partial x^\nu} - \frac{\partial\psi^*}{\partial x^\nu}\psi\right) \\ &\qquad + e^2\Phi_\nu\Phi_\nu\psi^*\psi + e^2\Phi_0^2\psi^*\psi \end{aligned}\right\}. \tag{21.11}$$

This is for plane surfaces, $t = \text{const.}$

The generalization to curved surfaces σ takes place by means of a normal vector N_ν:

$$N_\nu N_\nu = -1 \,.$$

For the plane surfaces $t=\text{const}$, $N_\nu=(0, 0, 0, i)$. Then,

$$\mathscr{H}_{\text{int}} = ie\Phi_\nu \left(\psi^* \frac{\partial\psi}{\partial x^\nu} - \frac{\partial\psi^*}{\partial x^\nu}\psi\right) + e^2\Phi_\nu\Phi_\nu\psi^*\psi + e^2(\Phi_\nu N_\nu)^2\psi^*\psi \,. \qquad [21.12]$$

Here $\mathscr{H}_{\text{int}}$ is, therefore, not invariant.

The auxiliary condition of electrodynamics

For the Heisenberg representation, we had Eq. [15.9],

$$\frac{\partial\Phi_\mu}{\partial x^\mu}\cdot\Psi = 0 \,.$$

This is, naturally, still true in the interaction representation at equal times:

$$\left(\frac{\partial\Phi_\mu}{\partial x^\mu}\right)_t \cdot\Psi(t) = 0 \,. \qquad [21.13]$$

If one chooses two different times, then an additional term which we wish to derive comes in.

If $\Box u = 0$, then

$$u(\boldsymbol{x}, t) = \int_{t''=t'} \mathrm{d}^3x'' \left[u(x'')\frac{\partial D(x-x'')}{\partial t''} - \frac{\partial u(x'')}{\partial t''} D(x-x'')\right]$$

is the solution with the boundary values

$$u(\boldsymbol{x}, t') \quad \text{and} \quad \left(\frac{\partial u(\boldsymbol{x}, t'')}{\partial t''}\right)_{t''=t'} \qquad [21.14]$$

at $t=t'$. This follows immediately from the properties

$$D(\boldsymbol{x}, 0) = 0 \quad \text{and} \quad \frac{\partial D}{\partial t}(\boldsymbol{x}, 0) = -\delta^3(\boldsymbol{x}) \,.$$

Thus,

$$\left(\frac{\partial \Phi_\mu}{\partial x^\mu}\right) = \int\limits_{t'=t_0} \mathrm{d}^3x' \left[\frac{\partial D(x-x')}{\partial t'} \frac{\partial \Phi_\mu(x')}{\partial x'^\mu} - D(x-x') \frac{\partial^2 \Phi_\mu(x')}{\partial x'^\mu \partial t'}\right],$$

since $\partial \Phi_\mu / \partial x^\mu$ in the interaction representation satisfies the force-free equations. Then,

$$\left(\frac{\partial \Phi_\mu}{\partial x^\mu}\right)_t \cdot \Psi(t_0) = -\int\limits_{t'=t_0} \mathrm{d}^3x' D(x-x') \frac{\partial^2 \Phi_\mu}{\partial x'^\mu \partial t'} \cdot \Psi(t_0),$$

since the first term above vanishes. Hence

$$\left(\frac{\partial \Phi_\mu}{\partial x^\mu}\right)_t \cdot \Psi(t_0) = -\left(\int\limits_{t'} \mathrm{d}^3x' D(x-x') \frac{\partial^2 \Phi_\mu(x')}{\partial x'^\mu \partial t'} \Psi(t')\right)_{t'=t_0}.$$

Upon partial integration, again because of the familiar auxiliary condition for equal times, only $\partial \psi / \partial t'$ contributes:

$$\left(\frac{\partial \Phi_\mu}{\partial x^\mu}\right)_t \cdot \Psi(t_0) = \left(\int\limits_{t'} \mathrm{d}^3x' D(x-x') \frac{\partial \Phi_\mu(x')}{\partial x'^\mu} [-iH_{\mathrm{int}}(t') \Psi(t')]\right)_{t'=t_0},$$

since

$$\frac{\partial \Psi}{\partial t} = -iH_{\mathrm{int}} \Psi.$$

Again using the auxiliary condition for equal times, this can be written as follows:

$$\left(\frac{\partial \Phi_\mu}{\partial x^\mu}\right)_t \cdot \Psi(t_0) = -i\left(\int\limits_{t'} \mathrm{d}^3x' D(x-x') \cdot \left[\frac{\partial \Phi_\mu(x')}{\partial x'^\mu}, H_{\mathrm{int}}(t')\right] \Psi(t')\right)_{t'=t_0},$$

$$\left(\frac{\partial \Phi_\mu}{\partial x^\mu}\right)_t \cdot \Psi(t_0) = -i\int\limits_{t'=t_0} \mathrm{d}^3x' D(x-x') \cdot \left[\frac{\partial \Phi_\mu(x')}{\partial x'^\mu}, H_{\mathrm{int}}(t')\right] \cdot \Psi(t_0). \qquad [21.15]$$

Now, we have already said that for spin-$\frac{1}{2}$ and spin-0,

$$\left[\frac{\partial \Phi_\mu(x')}{\partial x'^\mu}, \mathscr{H}_{\text{int}}(x'')\right]_{t'=t''} = ij^0(\boldsymbol{x}', t)\,\delta^3(\boldsymbol{x}' - \boldsymbol{x}'') . \qquad [21.16]$$

Thus,

$$\left[\frac{\partial \Phi_\mu(x')}{\partial x'^\mu}, H_{\text{int}}(t')\right] = ij^0(\boldsymbol{x}', t') ,$$

$$\left(\frac{\partial \Phi_\mu}{\partial x^\mu}\right)_t \cdot \Psi(t_0) = \int_{t'=t_0} \mathrm{d}^3x' D(x - x') j^0(\boldsymbol{x}', t_0)\, \Psi(t_0) ,$$

or

$$\left\{\frac{\partial \Phi_\mu(x)}{\partial x^\mu} - \int_{t'=t_0} D(x - x') j^0(\boldsymbol{x}', t_0)\, \mathrm{d}^3x'\right\} \Psi(t_0) = 0 . \qquad [21.17]$$

With curved surfaces, $j^0 = -j^\alpha N_\alpha$, and

$$\left\{\frac{\partial \Phi_\mu(x)}{\partial x^\mu} + \int_\sigma (j^\alpha N_\alpha) D(x - x')\, \mathrm{d}^3x'\right\} \Psi(\sigma) = 0 . \qquad [21.18]$$

Note: This is valid for spin-0 as well as spin-$\frac{1}{2}$.

22. DYSON'S INTEGRATION METHOD[4]

We are concerned with the integration of Eq. [21.5],

$$i\frac{\partial U}{\partial t} = H_{\text{int}}\, U,$$

for the operator U which transforms from the Heisenberg to the interaction representation. Here U also transforms from $\Psi(t_0)$ to $\Psi(t)$ in the interaction representation:

$$\frac{\partial \Psi_i}{\partial t} = -\,iH_{\text{int}}\Psi_i .$$

Thus,

$$\Psi_i(t) = U(t_0, t)\, \Psi_i(t_0) , \qquad [22.1]$$

[4] F. J. Dyson, *Phys. Rev.* **75**, 486, 1736 (1949).

where

$$\frac{\partial U(t_0, t)}{\partial t} = -\, iH_{\text{int}}(t)\, U(t_0, t)\ .$$

Dyson's formula: If P denotes the time-ordered product, then

$$U(t_0, t) = P\left(\exp\left[-i\int_{t_0}^{t} H_{\text{int}}(t')\, \mathrm{d}t'\right]\right) \qquad [22.2]$$

if $t \geqslant t_0$.

Note: 1. The exponential is symbolic and is meant to represent a series:

$$U(t_0, t) = \sum_{n=0}^{\infty} \frac{(-i)^n}{n!} \int_{t_0}^{t} \mathrm{d}t_1 \ldots \int_{t_0}^{t} \mathrm{d}t_n P(H_{\text{int}}(t_1) \ldots H_{\text{int}}(t_n))\ . \qquad [22.3]$$

2. Here P is defined as before by [15.24]:

$$P(A(t_1)B(t_2)) = \begin{cases} A(t_1)B(t_2) & (t_1 > t_2)\,, \\ B(t_2)A(t_1) & (t_1 < t_2)\ . \end{cases}$$

For $t_1 = t_2$, this is not defined.

3. Proof: It is almost obvious that $U(t_0, t)$ satisfies the differential equation of [21.5], because the $H_{\text{int}}(t_i)$ which is brought down upon differentiation contains the largest time, and, therefore, stands all the way to the left.

4. Conversely,

$$U^{-1}(t_0, t) = \sum_{n=0}^{\infty} \frac{(+i)^n}{n!} \int_{t_0}^{t} \mathrm{d}t_1 \ldots \int_{t_0}^{t} \mathrm{d}t_n P_-(H_{\text{int}}(t_1) \ldots H_{\text{int}}(t_n))\ ,$$

where P_- represents the reversed chronological ordering (largest time to the right).

5. The essential assumption in this is that

$$\lim_{\tau \to 0} \frac{1}{\tau} \int_{t}^{t+\tau} \mathrm{d}t_1 \int_{t}^{t+\tau} dt_2 P(H_{\text{int}}(t_1) H_{\text{int}}(t_2)) = 0\ ,$$

which is not obvious for a singular $P(H \cdot H)$. (See the case of spin-0 below.)

a. *Connection with the Heisenberg S-matrix*[5]

We define

$$S \equiv U(-\infty, +\infty)\,. \qquad [22.4]$$

This operator is different from zero only on the energy shell—*i.e.*, only for states of the same total energy.

In order to render this connection formally precise, we write

$$U(t_0, t) = 1 + \int_{t_0}^{t} W(t')\,\mathrm{d}t' = 1 + V(t_0, t)\,, \qquad [22.5]$$

where

$$W(t') = -\,iH_{\text{int}}(t') + \dots\,.$$

We now specialize to a representation in which, in the sense of Dirac, the states are characterized by the eigenvalues $q_0, q_1, q_2, \dots$ of a complete set of commuting operators. Here, the q's are, in addition, to be integrals of the system without interaction; that is,

$$[q, H_0] = 0\,.$$

Then

$$(q_1|\,W(t)|q_0) = \frac{1}{2\pi}\,(q_1|R|q_0)\cdot\exp\left[+\,i(\omega_1 - \omega_0)t\right]. \qquad [22.6]$$

Note: This is true only *cum grano salis*. In order that the initial state may be chosen as an eigenstate of H_0, the interaction must be switched on adiabatically (as, for example, if instead of H_{int} one writes $\exp[-\varepsilon|t|]H_{\text{int}}$, with very small ε). Instead of this, one can also choose as the initial state a suitable packet of eigenstates of H_0 which corresponds to a wave train bounded in time having the "length" $2T$. Then, for $|t| > T$, the interaction can be neglected. This wave packet does not cor-

[5] W. Heisenberg, *Z. Physik* **120**, 513, 673 (1943).

respond to a sharp value of H_0, and $(q_1|W(t)|q_0)$ is no longer rigorously monochromatic. Instead of $V(-\infty, t)$, one must then consider $V(-t_1, +t_1)$, where $t_1 > T$, so that at the times $-t_1$ and $+t_1$ the system is free of interactions. After this, one may let $t_1 \to \infty$, and only then let $T \to \infty$. In the form in which the calculation appears here, these two passages to the limit are reversed: We have, without leave, first let $T \to \infty$ and only then taken $t_1 \to \infty$. From this stems the paradox that

$$|(q_1|V(-\infty, t)|q_0)|^2$$

is independent of t. However, a rigorous calculation made in accordance with the above principle shows that the results are not changed thereby.

Then,

$$(q_1|V(-\infty, t)|q_0) = (q_1|R|q_0)\frac{1}{2\pi}\int_{-\infty}^{t} \exp[-i(\omega_0 - \omega_1)t']\,\mathrm{d}t' ,$$

or

$$(q_1|V(-\infty, t)|q_0) = (q_1|R|q_0)\exp[-i(\omega_0 - \omega_1)t]\delta_-(\omega_0 - \omega_1) . \qquad [22.7]$$

Note: In this,

$$\delta_\pm(\omega) = \frac{1}{2}\left[\delta(\omega) \pm \frac{1}{i\pi\omega}\right] = \frac{1}{2\pi}\int_0^\infty \exp[\mp i\omega t]\,\mathrm{d}t = \frac{1}{2\pi}\int_{-\infty}^0 \exp[\pm i\omega t]\,\mathrm{d}t .$$

For $t \to \infty$, there then follows

$$(q_1|V(-\infty, +\infty)|q_0) = (q_1|R|q_0)\,\delta(\omega_0 - \omega_1) . \qquad [22.8]$$

This is based on the following lemma:

$$\lim_{t\to+\infty}\int f(\omega)\exp[-i\omega t]\delta_+(\omega)\,\mathrm{d}\omega = 0 , \qquad [22.9]$$

where $f(\omega)$ is arbitrary but regular. Then, because $\delta_-(\omega) = \delta(\omega) - \delta_+(\omega)$, it follows immediately that in $V(-\infty, t)$, for $t \to \infty$ one may substitute $\delta_-(\omega) \to \delta(\omega)$ and $\exp[i\omega t] \to 1$.

We prove the lemma as follows:

$$J=\int f(\omega)\delta_+(\omega)\exp[-i\omega t]\,d\omega=\frac{1}{2\pi i}\int_{C_+}\frac{f(\omega)}{\omega}\exp[-i\omega t]\,d\omega\,,$$

where the integration path, C_+, is shown in Fig. 22.1. By

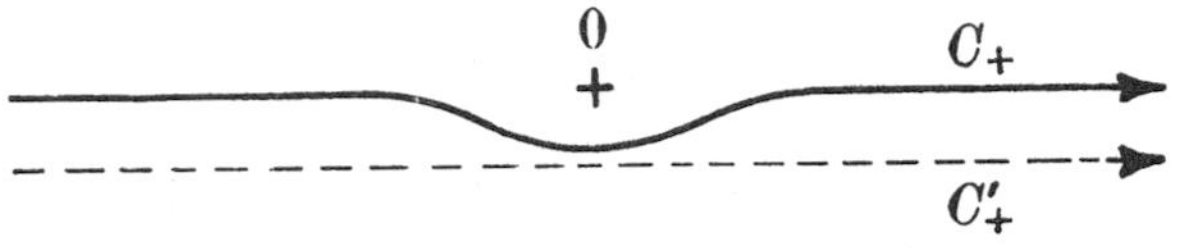

Figure 22.1

deforming C_+ into C'_+, we get

$$J=\frac{1}{2\pi i}\int_{-\infty}^{+\infty} f(x-i\varepsilon)\exp[-ixt-\varepsilon t]\frac{dx}{x-i\varepsilon}\to 0\,,\qquad \text{for } t\to\infty\,.$$

b. Transition probabilities

Above all, the S-matrix serves for the calculation of cross-sections. For this purpose, then, we consider the transition probability:

$$\left|\left(q_1\left|V\left(-\frac{T}{2},+\frac{T}{2}\right)\right|q_0\right)\right|^2=|(q_1|R|q_0)|^2\cdot\frac{1}{4\pi^2}\left|\int_{-T/2}^{+T/2}\exp[-i\omega t]\,dt\right|^2$$
$$=|(q_1|R|q_0)|^2\cdot\frac{1}{4\pi^2}\left|\frac{\sin(\omega T/2)}{(\omega/2)}\right|^2.$$

For large T $(T\gg 1/\omega)$, it is true that

$$\frac{1}{2\pi}\cdot\frac{\sin^2((\omega/2)T)}{(\omega/2)^2T}\to\delta(\omega)\,.$$

This is obvious; the normalization checks:

$$\frac{1}{2\pi}\int_{-\infty}^{+\infty}\frac{\sin^2((\omega/2)T)}{(\omega/2)^2T}\,d\omega=1\,.$$

With this, for large T,

$$\left|\left(q_1\left|V\left(-\frac{T}{2},+\frac{T}{2}\right)\right|q_0\right)\right|^2 \simeq \frac{1}{2\pi}|(q_1|R|q_0)|^2\cdot\delta(\omega)\cdot T\,.$$

Then, the transition probability per unit time is

$$W=\frac{1}{2\pi}|(q_1|R|q_0)|^2\cdot\delta(\omega)\,. \qquad [22.10]$$

We will use this result later.

Remark on the S-matrix: Heisenberg originally hoped to be able to make simple statements about the S-matrix for theories without a Hamiltonian formalism as well. However, it is now generally agreed that too little follows from a theory which contains only the S-matrix. In such a case, time no longer appears, and this is certainly going too far.

c. Additional remarks on Dyson's formula

Equation [22.3],

$$U(t_0,t)=\sum_{n=0}^{\infty}\frac{(-i)^n}{n!}\int_{t_0}^{t}\mathrm{d}t_1\ldots\int_{t_0}^{t}\mathrm{d}t_n P\big(H_{\mathrm{int}}(t_1)\ldots H_{\mathrm{int}}(t_n)\big)\,,$$

can be written as

$$U(t_0,t)=\sum_{n=0}^{\infty}\frac{(-i)^n}{n!}\int_{t_0}^{t}\mathrm{d}^4x_1\ldots\int_{t_0}^{t}\mathrm{d}^4x_n P\big(\mathscr{H}_{\mathrm{int}}(x_1)\ldots\mathscr{H}_{\mathrm{int}}(x_n)\big). \qquad [22.11]$$

Here, for spin-$\frac{1}{2}$, where $\mathscr{H}_{\mathrm{int}}$ is invariant, the Lorentz invariance becomes evident because the integrand is Lorentz invariant. (The volume of integration, however, is not but that is an inherent feature. Only for $U(-\infty,+\infty)\equiv S$ does the whole expression become invariant.)

Using the surfaces σ, one can write

$$U(\sigma_0,\sigma)=\sum_{n=0}^{\infty}\frac{(-i)^n}{n!}\int_{\sigma_0}^{\sigma}\mathrm{d}^4x_1\ldots$$

$$\int_{\sigma_n}^{\sigma}\mathrm{d}^4x_n P\big(\mathscr{H}_{\mathrm{int}}(x_1)\ldots\mathscr{H}_{\mathrm{int}}(x_n)\big)\,, \qquad [22.12]$$

where

$$\frac{\delta U(\sigma_0,\sigma)}{\delta\Omega(x)} = -\, i\mathscr{H}_{\text{int}}(x)\, U(\sigma_0,\sigma)\,. \qquad [22.13]$$

23. THE P^* PRODUCT FOR SPIN-ZERO

1. For the case of zero spin, there arises the already often mentioned complication that $\mathscr{H}_{\text{int}}(x)$ is not invariant and depends upon the surface normal. We have

$$\mathscr{H}_{\text{int}} = -\,\mathscr{L}_{\text{int}} + e^2(\Phi_\alpha N_\alpha)^2\psi^*\psi\,, \qquad [23.1]$$

where $\mathscr{L}_{\text{int}}$ is the Lagrangian density of the interaction,

$$\mathscr{L}_{\text{int}} = -\, ie\Phi_\mu\left(\psi^*\frac{\partial\psi}{\partial x^\mu} - \frac{\partial\psi^*}{\partial x^\mu}\psi\right) - e^2\Phi_\alpha\Phi_\alpha\psi^*\psi\,, \qquad [23.2]$$

and N_μ is the normal to σ at the point x.

2. The P symbol is undefined for equal times. This makes no difference as long as

$$[\mathscr{H}_{\text{int}}(\boldsymbol{x},t),\,\mathscr{H}_{\text{int}}(\boldsymbol{x}',t)] = 0$$

for equal times. For spin-zero, however,

$$[\mathscr{H}_{\text{int}}(x),\,\mathscr{H}_{\text{int}}(x')] = e^2\Phi_\mu(x)\,\Phi_\nu(x')\left\{\psi^*(x)\,\psi(x')\left[\frac{\partial\psi}{\partial x^\mu},\,\frac{\partial\psi'^*}{\partial x'^\nu}\right]\right.$$

$$\left.+\,\psi(x)\,\psi^*(x')\left[\frac{\partial\psi^*}{\partial x^\mu},\,\frac{\partial\psi'}{\partial x'^\nu}\right]\right\} + \begin{pmatrix}\text{terms which have weaker sin-}\\ \text{gularities at } t=t' \text{ and thus}\\ \text{do not contribute}\end{pmatrix},$$

$$[\mathscr{H}_{\text{int}}(x),\,\mathscr{H}_{\text{int}}(x')] = ie^2\Phi_\mu(x)\,\Phi_\nu(x')\left\{\psi^*(x)\,\psi(x')\,\frac{\partial^2\Delta(x-x')}{\partial x^\mu\partial x'^\nu}\right.$$

$$\left.+\,\psi(x)\,\psi^*(x')\,\frac{\partial^2\Delta(x-x')}{\partial x^\mu\partial x'^\nu}\right\} + \cdots\,, \qquad [23.3]$$

$$\left(\text{or also } = 2ie^2\Phi_\mu(x)\,\Phi_\nu(x')\,\psi^*(x)\,\psi(x')\,\frac{\partial^2\Delta(x-x')}{\partial x^\mu\partial x'^\nu} + \cdots\right).$$

Nishijima[6] has shown how these two difficulties can be circumvented if one introduces a somewhat modified form of the P-product. The Dyson P-product of two factors was (Eq. [15.25])

$$P(A(t)B(t')) = \tfrac{1}{2}\{A(t), B(t')\} + \tfrac{1}{2}\varepsilon(t-t')[A(t), B(t')],$$

which is undefined for $t=t'$. We define a modified product P^* in such a way that if $[A, B]$ contains a "critical" term of the type

$$K_{\mu\nu}\partial^2\Delta(x-x')/\partial x^\mu \partial x'^\nu$$

i.e., when

$$[A, B] = [A, B]_{\text{reg}} + K_{\mu\nu}\frac{\partial^2\Delta(x-x')}{\partial x^\mu \partial x'^\nu}, \qquad [23.4]$$

then we are always to have

$$P^*(A(x)\,B(x')) = \tfrac{1}{2}\{A(x), B(x')\} + \tfrac{1}{2}\,\varepsilon(x-x')[A(x), B(x')]_{\text{reg}} - K_{\mu\nu}\frac{\partial^2\bar{\Delta}(x-x')}{\partial x^\mu \partial x'^\nu}. \qquad [23.5]$$

That is, the ε on the right is always to be pulled through the differentiation.

Note: This is a modification only for $t=t'$; indeed only for $x=x'$.

A Lemma:

$$L \equiv \lim_{\tau\to 0}\frac{1}{\tau}\int_t^{t+\tau}dt'\int_t^{t+\tau}dt''\int d^3x''\,\frac{\partial^2\bar{\Delta}(x'-x'')}{\partial x'^\mu \partial x''^\nu} = \delta_{\mu 4}\delta_{\nu 4}, \qquad [23.6]$$

since

$$L = -\,\delta_{\mu 4}\delta_{\nu 4}\lim_{\tau\to 0}\frac{1}{\tau}\int_t^{t+\tau}dt'\int_t^{t+\tau}dt''\int d^3x''(\Box' - m^2)\bar{\Delta}(x'-x'')$$

$$= +\,\delta_{\mu 4}\delta_{\nu 4}\lim_{\tau\to 0}\frac{1}{\tau}\int_t^{t+\tau}dt'\int_t^{t+\tau}dt''\int d^3x''\delta^4(x'-x'') = +\,\delta_{\mu 4}\delta_{\nu 4}.$$

[6] K. Nishijima, *Prog. Theoret. Phys.* **5**, 405 (1950).

Then there follows, for example,

$$\lim_{\tau\to 0}\frac{1}{\tau}\int d^3x'\int d^3x''\int_t^{t+\tau} dt'\int_t^{t+\tau} dt''\, K_{\mu\nu}(x', x'')\,\frac{\partial^2\bar{\Delta}(x'-x'')}{\partial x'^{\mu}\partial x''^{\nu}}$$

$$=\int d^3x'\; K_{44}(\boldsymbol{x}', t;\, \boldsymbol{x}', t)\,. \qquad [23.7]$$

Now, it is to be shown that we obtain a correct solution of the equation

$$\frac{\partial U}{\partial t} = -i\int d^3x\; \mathscr{H}_{\text{int}}\cdot U$$

when we set

$$U = \sum_{n=0}^{\infty}\frac{(-i)^n}{n!}\int_{t_0}^{t} d^4x_1$$

$$\ldots\int_{t_0}^{t} d^4x_n\; P^*(-\mathscr{L}_{\text{int}}(x_1)\,\ldots,\, -\mathscr{L}_{\text{int}}(x_n))\,, \qquad [23.8]$$

where $\mathscr{L}_{\text{int}}(x)$ represents the invariant interaction Lagrangian density.

Note: 1. We had

$$\mathscr{H}_{\text{int}}(x) = -\mathscr{L}_{\text{int}}(x) + e^2\Phi_0^2(x)\,\psi^*(x)\,\psi(x)\,.$$

Furthermore,

$$[\mathscr{H}_{\text{int}}(x), \mathscr{H}_{\text{int}}(x')] = 2ie^2\,\Phi_\mu(x)\,\Phi_\nu(x)\,\psi^*(x)\,\psi(x)\,\frac{\partial^2\Delta(x-x')}{\partial x^\mu\partial x'^\nu} + \ldots$$

$$= K_{\mu\nu}(x)\,\frac{\partial^2\Delta(x-x')}{\partial x^\mu\partial x'^\nu}\,,$$

from which

$$\mathscr{H}_{\text{int}}(x) = -\mathscr{L}_{\text{int}}(x) + \frac{i}{2}K_{44}(x)\,. \qquad [23.9]$$

2. The critical terms in the commutation relations come from the commutator

$$\left[\frac{\partial\psi^*}{\partial x^\mu}, \frac{\partial\psi'}{\partial x'^\nu}\right]$$

and thus from the term $[\mathscr{L}_{\text{int}}, \mathscr{L}_{\text{int}}]$ in $[\mathscr{H}_{\text{int}}, \mathscr{H}_{\text{int}}]$ and not from the other term. That is,

$$[-\mathscr{L}_{\text{int}}(x), -\mathscr{L}_{\text{int}}(x')] = K_{\mu\nu}(x)\frac{\partial^2 \Delta(x-x')}{\partial x^\mu \partial x'^\nu} + \dots . \qquad [23.10]$$

We now calculate $\partial U/\partial t$:

$$\frac{\partial U}{\partial t} = \lim_{\tau\to 0}\frac{1}{\tau}\left(U(t+\tau) - U(t)\right).$$

Besides the trivial term $-\int d^3x \mathscr{L}_{\text{int}} \cdot U$, there is still an additional term from the double integrals (see the Lemma). Higher (triple and higher) integrals contribute nothing as long as at most second derivatives of $\bar{\Delta}$ appear.

Since there are $\binom{n}{2}$ pairs of integrals, exactly the right combinatorial factor is obtained. Thus,

$$\frac{\partial U}{\partial t} = (-i)\left(-\int d^3x\, \mathscr{L}_{\text{int}}\right)\cdot U + \frac{(-i)^2}{2!}\lim_{\tau\to 0}\frac{1}{\tau}\int d^3x' \int_t^{t+\tau} dt' \int d^3x'' \int_t^{t+\tau} dt''$$
$$\cdot P^*\left(\mathscr{L}_{\text{int}}(x')\, \mathscr{L}_{\text{int}}(x'')\right)\cdot U,$$

in which

$$P^*\left(\mathscr{L}_{\text{int}}(x')\, \mathscr{L}_{\text{int}}(x'')\right) \to - K_{\mu\nu}(x')\frac{\partial^2 \bar{\Delta}(x'-x'')}{\partial x'^\mu \partial x''^\nu}.$$

Thus, with the Lemma of Eq. [23.7],

$$\frac{\partial U}{\partial t} = (-i)\left(-\int d^3x \mathscr{L}_{\text{int}}\right)\cdot U + \frac{1}{2}\int d^3x'\, K_{44}(x')\cdot U,$$

$$i\frac{\partial U}{\partial t} = -\int d^3x\left(\mathscr{L}_{\text{int}}(x) - \frac{i}{2}K_{44}(x)\right)\cdot U,$$

so that finally

$$i\frac{\partial U}{\partial t} = H_{\text{int}}\, U. \qquad \text{Q.E.D.}$$

Remark: All of the above is based on the fact that the critical term in the commutator and the noninvariant term in the Hamiltonian are the same. This is not accidental, and is quite generally valid.

As the definition of $K_{\mu\nu}(x)$, let

$$[\mathscr{H}_{\text{int}}(x), \mathscr{H}_{\text{int}}(x')] = K_{\mu\nu}(x)\,\frac{\partial^2 \Delta(x-x')}{\partial x^\mu \partial x'^\nu} + \text{noncritical terms}\,. \qquad [23.11]$$

Then,

$$\mathscr{H}_{\text{int}}(x) = -\,\mathscr{L}_{\text{int}}(x) - \frac{i}{2} K_{\mu\nu} N_\mu N_\nu\,, \qquad N_\mu N_\mu = -1\,. \qquad [23.12]$$

Let N be in the 4 direction. Then, $N_4 = i$, and

$$\mathscr{H}_{\text{int}}(x) = -\,\mathscr{L}_{\text{int}}(x) + \frac{i}{2} K_{44}(x)\,. \qquad [23.13]$$

We now wish to prove this relationship generally. That is, we want to show that this relationship is always valid whenever the interaction Lagrangian density is at most linear in the derivatives of the fields.

The equation of motion for the state $\Psi(\sigma)$ is

$$\left(\mathscr{H}_{\text{int}}(x, N) - i\,\frac{\delta}{\delta\Omega(x)}\right)\Psi(\sigma) = 0\,. \qquad [23.14]$$

In order that this equation be satisfied, the following integrability condition must be valid:

$$\left[\mathscr{H}_{\text{int}}(x, N) - i\,\frac{\delta}{\delta\Omega(x)}, \mathscr{H}_{\text{int}}(x', N') - i\,\frac{\delta}{\delta\Omega(x')}\right] = 0\,. \qquad [23.15]$$

Since all such theories are, indeed, derived from a self-consistent formulation in the Heisenberg representation, this requirement is always satisfied.

Then, it follows that

$$[\mathscr{H}_{\text{int}}(x, N), \mathscr{H}_{\text{int}}(x', N')] = i\,\frac{\delta}{\delta\Omega(x)}\,\mathscr{H}_{\text{int}}(x', N') - i\,\frac{\delta}{\delta\Omega(x')}\,\mathscr{H}_{\text{int}}(x, N)\,. \qquad [23.16]$$

If we now consider the critical terms, we can antisymmetrize with respect to x and x', and we obtain on the left

$$\frac{1}{2}\left(K_{\mu\nu}(x)\frac{\partial^2\Delta(x-x')}{\partial x^\mu \partial x'^\nu} - K_{\mu\nu}(x')\frac{\partial^2\Delta(x'-\tau)}{\partial x'^\mu \partial x^\nu}\right).$$

Equation [23.16] is then satisfied if we set

$$i\frac{\delta}{\delta\Omega(x)}\mathscr{H}_{\text{int}}(x', N') = \frac{1}{2}K_{\mu\nu}(x)\frac{\partial^2\Delta(x-x')}{\partial x^\mu \partial x'^\nu}. \qquad [23.17]$$

We need Gauss's theorem:

$$-\oint_{\sigma'} F_\mu N_\mu \, d^3\sigma = \int_\Omega \frac{\partial F_\mu}{\partial x^\mu} d^4x. \qquad [23.18]$$

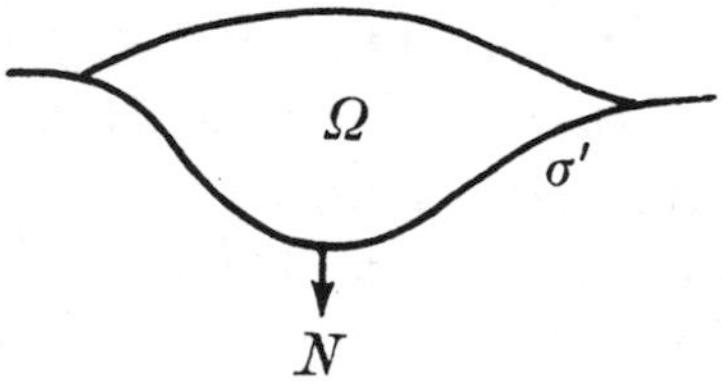

Figure 23.1

That is,

$$-\frac{\delta}{\delta\Omega(x)}\int_\sigma F_\mu N_\mu \, d^3\sigma = \frac{\partial F_\mu}{\partial x^\mu}. \qquad [23.19]$$

With this, we can integrate Eq. [23.17]:

$$\mathscr{H}_{\text{int}}(x', N') = \frac{i}{2}\int_\sigma K_{\mu\nu}(x)\frac{\partial\Delta(x-x')}{\partial x'^\nu}N_\mu \, d^3\sigma. \qquad [23.20]$$

Note: Strictly speaking, one should somehow regularize the singular Δ in order that the transformation be possible.

If we specialize the coordinate system so that $N_4 = i$, then

$$N_4 = i, \quad \mu = 4, \quad \frac{\partial\Delta(x-x')}{\partial x'^4} = -i\frac{\partial\Delta}{\partial x'^0} = -i\delta^3(\boldsymbol{x}-\boldsymbol{x}').$$

Thus,

$$\mathscr{H}_{\text{int}}(x', N') = +\frac{i}{2}K_{44}(x') = -\frac{i}{2}K_{\mu\nu}(x')N'_\mu N'_\nu\,, \qquad [23.21]$$

and we have thus found the relation which we have sought. With this, we have also shown that the P^* symbol allows the terms which depend on the normal to be circumvented, not only for the scalar theory, but also for all other theories (such as the vector theory) in which $\mathscr{L}_{\text{int}}$ contains at most linear terms in the derivatives of the field quantities.

Chapter 5. Heisenberg Representation: S-Matrix and Charge Renormalization

24. THE S-MATRIX AND THE HEISENBERG REPRESENTATION [1]

The S-matrix can be defined without reference to the surfaces, although Dyson's formula remains simpler for practical calculations.

In the Heisenberg representation, we have

$$\begin{cases} \left(\gamma^\nu \dfrac{\partial}{\partial x^\nu} + m\right)\psi = +ie\gamma^\nu \Phi_\nu \psi\,, \\ \dfrac{\partial \bar{\psi}}{\partial x^\nu}\gamma^\nu - m\bar{\psi} = -ie\bar{\psi}\gamma^\nu \Phi_\nu\,, \\ \Box \Phi_\nu = -j^\nu\,, \qquad j^\nu = \dfrac{ie}{2}[\bar{\psi}, \gamma^\nu \psi] = ie\bar{\psi}\gamma^\nu\psi\,, \\ \dfrac{\partial \Phi_\nu}{\partial x^\nu}\cdot \Psi = 0. \end{cases}$$

(Compare Secs. 15 and 19.) We define incoming and outgoing fields by the equations

$$\left.\begin{aligned} \Phi_\nu(x) &= \Phi_\nu^{\text{in}}(x) + \int D^{\text{ret}}(x-x')\, j^\nu(x')\, \mathrm{d}^4x' \\ \psi(x) &= \psi^{\text{in}}(x) - ie\int S^{\text{ret}}(x-x')\gamma^\nu \Phi_\nu(x')\psi(x')\,\mathrm{d}^4x' \\ \bar{\psi}(x) &= \bar{\psi}^{\text{in}}(x) - ie\int \bar{\psi}(x')\gamma^\nu\Phi_\nu(x')\, S^{\text{adv}}(x'-x)\,\mathrm{d}^4x' \end{aligned}\right\}. \qquad [24.1]$$

[1] C. N. Yang and D. Feldman, *Phys. Rev.* **79**, 972 (1950).

For $t \to -\infty$, the integrals vanish in accordance with the definitions of the retarded functions, so that the incoming fields become identical with the complete fields. For $t \to -\infty$, the free commutation relations become asymptotically valid for the Heisenberg fields. In addition, the incoming fields satisfy the free field equations,

$$\left.\begin{aligned} \left(\gamma^\nu \frac{\partial}{\partial x^\nu} + m\right) \psi^{\text{in}} &= 0 \\ \bar{\psi}^{\text{in}} \left(\gamma^\nu \frac{\overleftarrow{\partial}}{\partial x^\nu} - m\right) &= 0 \\ \Box \Phi_\nu^{\text{in}} &= 0 \end{aligned}\right\} . \qquad [24.2]$$

Thus, for the incoming fields the free commutation relations are valid:

$$\left.\begin{aligned} \{\psi_\alpha^{\text{in}}(x),\, \bar{\psi}_\beta^{\text{in}}(x')\} &= -\,i\, S_{\alpha\beta}(x - x') \\ [\Phi_\mu^{\text{in}}(x),\, \Phi_\nu^{\text{in}}(x')] &= i\delta_{\mu\nu} D(x - x') \end{aligned}\right\} . \qquad [24.3]$$

Analogously, we now define the outgoing fields:

$$\left.\begin{aligned} \Phi_\nu(x) &= \Phi_\nu^{\text{out}}(x) + \int D^{\text{adv}}(x - x')\, j^\nu(x')\, \mathrm{d}^4x' \\ \psi(x) &= \psi^{\text{out}}(x) - ie \int S_{\text{adv}}(x - x')\, \gamma^\nu \Phi_\nu(x')\, \psi(x')\, \mathrm{d}^4x' \\ \bar{\psi}(x) &= \bar{\psi}^{\text{out}}(x) - ie \int \bar{\psi}(x')\, \gamma^\nu \Phi_\nu(x')\, S^{\text{ret}}(x' - x)\, \mathrm{d}^4x' \end{aligned}\right\} . \qquad [24.4]$$

In exactly the same manner, it follows that these fields also satisfy the free field equations and the free commutation relations.

Since both sets of fields satisfy the same relations, there must exist a canonical transformation which relates them. Let $F = \psi$, $\bar{\psi}$ or Φ_ν, so that

$$F^{\text{out}}(x) = S^{-1} F^{\text{in}}(x) S\,. \qquad [24.5]$$

Here S is the S-matrix as it must be defined in the Heisenberg representation.

From this definition, one can, in principle, calculate the S-matrix recursively (in powers of e). However, this is much more complicated than in the Dyson formalism. We will not go into this further. Nevertheless, following R. Glauber [A-6], we will present the connection with the interaction representation. For this purpose we define new Δ functions which depend upon a surface σ:

$$\Delta^{\sigma}(x; x') = \begin{Bmatrix} \Delta^{\text{ret}}(x - x'), & \text{if } x' \text{ is later than } \sigma \\ \Delta^{\text{adv}}(x - x'), & \text{if } x' \text{ is earlier than } \sigma \end{Bmatrix}. \qquad [24.6]$$

Upon using

$$\varepsilon(x - x') \equiv \begin{Bmatrix} +1 & (t > t') \\ -1 & (t < t') \end{Bmatrix},$$

and, correspondingly,

$$\varepsilon(\sigma, x') \equiv \begin{cases} +1 & (x' \text{ earlier than } \sigma) \\ -1 & (x' \text{ later than } \sigma), \end{cases}$$

we have

$$\Delta^{\sigma}(x; x') = \frac{1 - \varepsilon(\sigma, x')}{2} \Delta^{\text{ret}}(x - x') + \frac{1 + \varepsilon(\sigma, x')}{2} \Delta^{\text{adv}}(x - x'),$$

or

$$\Delta^{\sigma}(x; x') = \frac{\varepsilon(\sigma, x') - \varepsilon(x - x')}{2} \Delta(x - x'). \qquad [24.7]$$

The dependence upon x' is, indeed, complicated; the dependence upon x is, however, the same as in $\bar{\Delta}(x - x')$. Thus,

$$(\square - m^2)\Delta^{\sigma}(x; x') = -\delta^4(x - x'). \qquad [24.8]$$

If x lies on σ (we write $x \subset \sigma$), then

$$\left. \begin{aligned} \Delta^{\sigma}(x; x')\big|_{x \subset \sigma} &= 0 \\ \left. \frac{\partial \Delta^{\sigma}(x; x')}{\partial x^{\mu}} \right|_{x \subset \sigma} &= 0 \end{aligned} \right\}. \qquad [24.9]$$

Not all of the second derivatives vanish. The 4, 4 component contributes, while the others do not. We obtain

$$\left.\frac{\partial^2 \Delta^\sigma(x;x')}{\partial x^4 \partial x^4}\right|_{x\subset\sigma} = (\Box - m^2)\Delta^\sigma(x,\sigma')|_{x\subset\sigma} = -\delta^4(x-x')\,,$$

so that

$$\left.\frac{\partial^2 \Delta^\sigma(x;x')}{\partial x^\mu \partial x^\nu}\right|_{x\subset\sigma} = +N_\mu N_\nu \delta^4(x-x')\,. \qquad [24.10]$$

Furthermore, we need

$$\frac{\delta\Delta^\sigma(x;x')}{\delta\Omega(y)} = \delta^4(x'-y)\,\Delta(x-x')\,. \qquad [24.11]$$

This follows immediately from the formula

$$\frac{\delta\varepsilon(\sigma,x')}{\delta\Omega(y)} = 2\delta^4(x'-y)\,, \qquad [24.12]$$

which is obvious from Fig. 24.1.

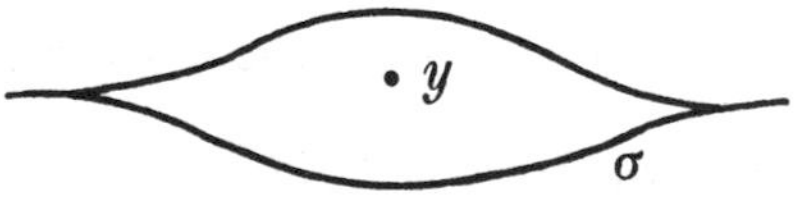

Figure 24.1

Clearly, we must then define

$$\left.\begin{aligned} S^\sigma(x;x') &= \left(\gamma\frac{\partial}{\partial x} - m\right)\Delta^\sigma(x;x') \\ D^\sigma(x;x') &= \Delta^\sigma(x;x')|_{m=0} \end{aligned}\right\}. \qquad [24.13]$$

With the aid of these functions we can, in addition to the incoming and outgoing fields, define further solutions of the free field equations which depend upon an arbitrary space-like surface σ:

$$\left.\begin{aligned} \psi(x) &= \psi(x,\sigma) - ie\int S^\sigma(x;x')\gamma^\nu\Phi_\nu(x')\psi(x')\,\mathrm{d}^4x' \\ \Phi_\nu(x) &= \Phi_\nu(x,\sigma) + \int D^\sigma(x;x')\,j^\nu(x')\,\mathrm{d}^4x' \end{aligned}\right\}. \qquad [24.14]$$

Obviously, it is true that

$$\left.\begin{aligned}&\left(\gamma\frac{\partial}{\partial x}+m\right)\psi(x,\sigma)=0\\&\Box\Phi_\nu(x,\sigma)=0\end{aligned}\right\}\qquad[24.15]$$

(for fixed σ).

For $x\subset\sigma$, it follows from the above properties of the Δ^σ that

$$\left.\begin{aligned}\psi(x,\sigma)|_{x\subset\sigma}&=\psi(x)|_{x\subset\sigma}\\\Phi_\nu(x,\sigma)|_{x\subset\sigma}&=\Phi_\nu(x)|_{x\subset\sigma}\end{aligned}\right\}.\qquad[24.16]$$

Furthermore, analogous to the interaction representation, we get

$$\frac{\partial\Phi_\nu(x)}{\partial x^\nu}\cdot\Psi=\left(\frac{\partial\Phi_\nu(x,\sigma)}{\partial x^\nu}-\int_\sigma d\sigma'^\nu j^\nu(x')D(x-x')\right)\Psi=0\,,$$

$$(d\sigma^\nu=N_\nu\cdot d\sigma)\,.\qquad[24.17]$$

In addition, the force-free commutation relations hold:

$$\left.\begin{aligned}\{\psi_\alpha(x,\sigma),\bar\psi_\beta(x',\sigma)\}&=-iS_{\alpha\beta}(x-x')\\ [\Phi_\mu(x,\sigma),\Phi_\nu(x',\sigma)]&=+i\delta_{\mu\nu}D(x-x')\end{aligned}\right\}.\qquad[24.18]$$

Note: For $\sigma\to+\infty$, $F(x,\sigma)\to F^{\text{out}}(x)$;

For $\sigma\to-\infty$, $F(x,\sigma)\to F^{\text{in}}(x)$.

Just as before, one now concludes that there exists a unitary transformation, $U(\sigma,\sigma')$:

$$F(x,\sigma)=U^{-1}(\sigma,\sigma')F(x,\sigma')U(\sigma,\sigma')\,.\qquad[24.19]$$

Here, we have the special case

$$U(+\infty,-\infty)=S\,.\qquad[24.20]$$

It is essential that the multiplication property

$$U(\sigma,\sigma')=U(\sigma'',\sigma')U(\sigma,\sigma'')\qquad[24.21]$$

holds. From this, it follows immediately that

$$\frac{\delta U(\sigma, \sigma')}{\delta \Omega(x)} = U(\sigma, \sigma') \frac{\delta U(\sigma, \sigma)}{\delta \Omega(x)} \qquad (x \subset \sigma) ,$$

where the precise meaning of $\delta U(\sigma, \sigma)/\delta \Omega(x)$ is

$$\lim_{\delta\sigma \to 0} \frac{U(\sigma + \delta\sigma(x), \sigma) - U(\sigma, \sigma)}{\delta \Omega(\sigma, x)} .$$

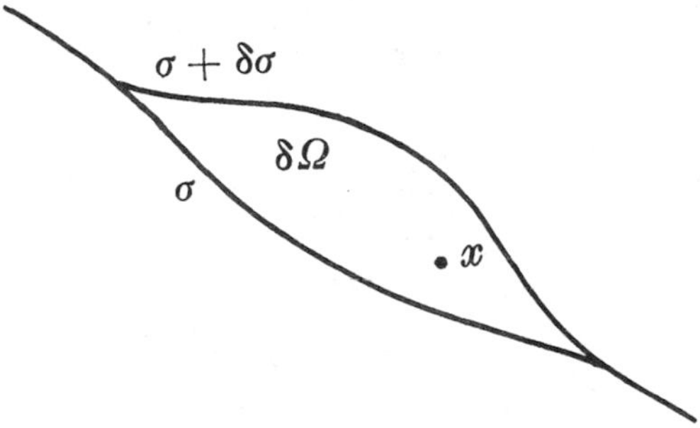

Figure 24.2

If we write, as a definition,

$$\frac{\delta U(\sigma, \sigma)}{\delta \Omega(x)} \equiv -i\mathscr{H}_{\text{int}}(x, \sigma)|_{x \subset \sigma} , \qquad [24.23]$$

then it follows that

$$\frac{\delta U(\sigma, \sigma')}{\delta \Omega(x)} = -iU(\sigma, \sigma')\mathscr{H}_{\text{int}}(x, \sigma)|_{x \subset \sigma} .$$

Also, because of Eq. [24.19],

$$\frac{\delta U(\sigma, \sigma')}{\delta \Omega(x)} = -i\mathscr{H}_{\text{int}}(x, \sigma')U(\sigma, \sigma') \qquad (x \subset \sigma) . \qquad [24.24]$$

We define $\mathscr{H}_{\text{int}}(x, -\infty) \equiv \mathscr{H}_{\text{int}}(x)$. Furthermore, it follows from Eq. [24.19] that

$$i \frac{\delta F(x, \sigma)}{\delta \Omega(x')} = [F(x, \sigma), \mathscr{H}_{\text{int}}(x', \sigma)]|_{x' \subset \sigma} . \qquad [24.25]$$

In order to establish the connection with the old theory, we must now show that the $\mathscr{H}_{\text{int}}$ here is the same as the previous one. From Eq. [24.14],

$$\left.\begin{aligned}\psi(x) &= \psi(x,\sigma) - ie\int S^\sigma(x;x')\gamma^\nu \Phi_\nu(x')\psi(x')\,\mathrm{d}^4x' \\ \Phi_\nu(x) &= \Phi_\nu(x,\sigma) + \int D^\sigma(x;x')\,j^\nu(x')\,\mathrm{d}^4x'\end{aligned}\right\}$$

one finds, by applying $\delta/\delta\Omega(x')$, that

$$\frac{\delta\psi(x,\sigma)}{\delta\Omega(x')} = ie\int \frac{\delta S^\sigma(x;x'')}{\delta\Omega(x')}\gamma^\nu \Phi_\nu(x'')\psi(x'')\,\mathrm{d}^4x'' ,$$

$$\frac{\delta\Phi_\nu(x,\sigma)}{\delta\Omega(x')} = -\int \frac{\delta D^\sigma(x;x'')}{\delta\Omega(x')}\,j^\nu(x')\,\mathrm{d}^4x'' .$$

Now, since

$$\frac{\delta S^\sigma(x;x'')}{\delta\Omega(x')} = \delta^4(x'-x'')\,S(x-x') ,$$

$$\frac{\delta D^\sigma(x;x'')}{\delta\Omega(x')} = \delta^4(x'-x'')\,D(x-x') ,$$

as we have seen above, it follows that

$$\begin{aligned}[\psi(x,\sigma), \mathscr{H}_{\mathrm{int}}(x',\sigma)]|_{x'\subset\sigma} &= -\,eS(x-x')\gamma^\nu\Phi_\nu(x')\psi(x') \\ &= -\,e\,S(x-x')\gamma^\nu\Phi_\nu(x',\sigma)\psi(x',\sigma)|_{x'\subset\sigma} ,\end{aligned}$$

$$\begin{aligned}[\Phi_\nu(x,\sigma), \mathscr{H}_{\mathrm{int}}(x',\sigma)]|_{x'\subset\sigma} &= -\,iD(x-x')\,j^\nu(x') \\ &= -\,i\,D(x-x')\,j^\nu(x',\sigma)|_{x'\subset\sigma} ,\end{aligned}$$

in which we have also used Eq. [24.16]. Here $\mathscr{H}_{\mathrm{int}}$ is uniquely determined to within a c-number; thus, U is determined to within a c-number phase factor, which does not interest us. The solution is

$$\mathscr{H}_{\mathrm{int}}(x',\sigma)|_{x'\subset\sigma} = -\,j^\nu(x',\sigma)\,\Phi_\nu(x',\sigma)|_{x'\subset\sigma} ,$$

and so

$$\mathscr{H}_{\mathrm{int}}(x,\sigma) = -\,j^\nu(x,\sigma)\,\Phi_\nu(x,\sigma) .$$

A special case is

$$\mathscr{H}_{\mathrm{int}}(x) = -\,j^{\nu\,\mathrm{int}}(x)\,\Phi_\nu^{\mathrm{int}}(x) . \qquad [24.26]$$

Yang-Feldman formalism for spin-0 particles

The field equations are

$$(\Box - m^2)\psi(x) = +ie\left[\frac{\partial}{\partial x^\mu}\big(\Phi_\mu(x)\psi(x)\big) + \Phi_\mu(x)\frac{\partial\psi(x)}{\partial x^\mu}\right] + e^2\Phi_\mu(x)\Phi_\mu(x)\psi(x)\,,$$

$$\Box\Phi_\mu(x) = -j^\mu(x) = -ie\left(\frac{\partial\psi^*(x)}{\partial x^\mu}\psi(x) - \psi^*(x)\frac{\partial\psi(x)}{\partial x^\mu}\right) + 2e^2\Phi_\mu(x)\psi^*(x)\psi(x)\,.$$

We then define

$$\left.\begin{aligned}\psi(x) = \psi(x,\sigma) - ie\bigg\{&\frac{\partial}{\partial x^\nu}\int\Delta^\sigma(x;x')\Phi_\nu(x')\psi(x')\,\mathrm{d}^4x' \\ &+\int\Phi_\nu(x')\frac{\partial\psi(x')}{\partial x'^\nu}\Delta^\sigma(x;x')\,\mathrm{d}^4x'\bigg\} \\ &- e^2\int\Phi_\nu(x')\Phi_\nu(x')\psi(x')\Delta^\sigma(x;x')\,\mathrm{d}^4x' \\ \Phi_\nu(x) = \Phi_\nu(x,\sigma) + &\int D^\sigma(x;x')j^\nu(x')\,\mathrm{d}^4x'\end{aligned}\right\}, \qquad [24.27]$$

(as before, but with a new j^μ). Then,

$$\left.\begin{aligned}(\Box - m^2)\psi(x,\sigma) &= 0 \\ \Box\Phi_\mu(x,\sigma) &= 0\end{aligned}\right\}. \qquad [24.28]$$

Furthermore, since the first derivatives of $\Delta^\sigma(x;x')$ vanish for $x\subset\sigma$ while the second derivatives do not, one sees immediately that

$$\left\{\begin{aligned}&\psi(x)|_{x\subset\sigma} = \psi(x,\sigma)|_{x\subset\sigma}\,, \\ &\Phi_\nu(x)|_{x\subset\sigma} = \Phi_\nu(x,\sigma)|_{x\subset\sigma}\,, \\ &\left.\frac{\partial\psi(x)}{\partial x^\nu}\right|_{x\subset\sigma} = \left.\frac{\partial\psi(x,\sigma)}{\partial x^\nu}\right|_{x\subset\sigma} - ieN_\mu N_\nu\Phi_\nu(x,\sigma)\psi(x,\sigma)|_{x\subset\sigma}\,.\end{aligned}\right.$$

This has a simple meaning. For surfaces $t = \text{const}$, we have

$$\left.\frac{\partial\psi(x)}{\partial \boldsymbol{x}}\right|_{x\subset\sigma}=\left.\frac{\partial\psi(x,\sigma)}{\partial \boldsymbol{x}}\right|_{x\subset\sigma},$$

$$\left.\frac{\partial\psi(x,\sigma)}{\partial t}\right|_{x\subset\sigma}=\left.\frac{\partial\psi(x)}{\partial t}\right|_{x\subset\sigma}+ie\Phi_0(x)\psi(x)|_{x\subset\sigma}=\pi^*(x)|_{x\subset\sigma}.$$

This must be true physically because the $\psi(x, \sigma)$'s must, indeed, satisfy the force-free field equations and commutation relations; therefore, the derivatives of the $\psi(x, \sigma)$'s must, at constant time, reduce to the π's.

Then, as before, we get for spin-$\frac{1}{2}$,

$$\begin{aligned}\frac{\delta\psi(x,\sigma)}{\delta\Omega(x')}=ie&\left\{\frac{\partial\Delta(x-x')}{\partial x^\nu}\Phi_\nu(x',\sigma)\psi(x',\sigma)\right.\\&\left.\left.+\Delta(x-x')\Phi_\nu(x',\sigma)\frac{\partial\psi(x',\sigma)}{\partial x'_\nu}\right\}\right|_{x'\subset\sigma}\\&+e^2\Phi_\nu(x',\sigma)\Phi_\nu(x',\sigma)\psi(x',\sigma)\Delta(x-x')|_{x'\subset\sigma}\\&+e^2(\Phi_\nu(x',\sigma)N^\nu)^2\psi(x',\sigma)\Delta(x-x')|_{x'\subset\sigma}.\end{aligned}$$

With

$$i\,\frac{\delta\psi(x,\sigma)}{\delta\Omega(x')}=[\psi(x,\sigma),\mathscr{H}(x',\sigma)]|_{x'\subset\sigma},$$

there follows

$$\begin{aligned}\mathscr{H}_{\text{int}}(x',\sigma)=ie\Phi_\nu(x',\sigma)&\left[\psi^*(x',\sigma)\frac{\partial\psi(x',\sigma)}{\partial x'^\nu}-\frac{\partial\psi^*(x',\sigma)}{\partial x'^\nu}\psi(x',\sigma)\right]\\&+e^2\Phi_\nu(x',\sigma)\Phi_\nu(x',\sigma)\psi^*(x',\sigma)\psi(x',\sigma)\\&+e^2(N_\nu\Phi_\nu(x',\sigma))^2\psi^*(x',\sigma)\psi(x',\sigma)\,.\end{aligned}\qquad[24.29]$$

We thus again obtain the expression for the interaction representation.

25. RENORMALIZED FIELDS IN THE HEISENBERG REPRESENTATION

In Eq. [11.15] it was shown that, because of the coupling to the photon field, all currents are multiplied by a factor

$$1+\gamma \equiv \frac{e}{e_0}, \quad \text{where} \quad \gamma = \frac{\alpha}{3\pi} \log \frac{m}{M}.$$

(We have already noted that there exists a question of definition here, with respect to the magnitude of this factor. According to Schwinger's method, one obtains a value of twice the above one (Feynman's). We will continue here with the Feynman value.)

We had shown that the additional current is (see Eq. [8.3])

$$j^{\mu\,\mathrm{pol}} = \frac{i}{2} \int \langle [j^\mu(x), j^\nu(x')] \rangle_0 \varepsilon(x-x')\, \mathscr{A}_\nu(x')\, d^4x',$$

and, with Eq. [7.14],

$$\frac{i}{2} \langle [j^\mu(x), j^\nu(x')] \rangle_0 \varepsilon(x-x') = -e^2 \overline{K}_{\mu\nu}(x-x'),$$

that

$$j^{\mu\,\mathrm{pol}} = -e^2 \int \overline{K}_{\mu\nu}(x-x')\, \mathscr{A}_\nu(x')\, d^4x'. \qquad [25.1]$$

Thus, with Eq. [11.7],

$$j^{\mu\,\mathrm{pol}} = 2\gamma j^\mu + c_1 \Box j^\mu + c_2 \Box\Box j^\mu + \dots. \qquad [25.2]$$

Now, it is to be expected that just as the currents, all fields are also multiplied by $(1+\gamma)$. We then consider the "renormalized fields,"

$$\Phi_\mu^R(x) = \frac{1}{1+\gamma} \Phi_\mu(x) = (1-\gamma+\dots)\Phi_\mu(x). \qquad [25.3]$$

For the nonrenormalized fields, with or without interaction, the canonical commutation relations,

$$i[\dot{\boldsymbol{\Phi}}_\mu(\boldsymbol{x}, t), \boldsymbol{\Phi}_\nu(\boldsymbol{x}', t)] = \delta_{\mu\nu}\delta^3(\boldsymbol{x}-\boldsymbol{x}'), \qquad [25.4]$$

hold. This is the "strong," nongauge-invariant form. If one wants gauge-invariant expressions, then one can consider

$$i[E_x(\boldsymbol{x}, t), H_y(\boldsymbol{x}', t)] = \frac{\partial}{\partial z}\delta^3(\boldsymbol{x}-\boldsymbol{x}') \quad \text{(and cyclical permutations).} \tag{25.5}$$

Now, following Jost (unpublished), we will consider the commutation relations for the renormalized fields:

$$\begin{aligned} i[\dot{\Phi}^R_\mu(\boldsymbol{x}, t), \Phi^R_\nu(\boldsymbol{x}', t)] &= \frac{1}{(1+\gamma)^2}\delta_{\mu\nu}\delta^3(\boldsymbol{x}-\boldsymbol{x}') \\ &= (1-2\gamma+\dots)\delta_{\mu\nu}\delta^3(\boldsymbol{x}-\boldsymbol{x}'), \end{aligned} \tag{25.6}$$

or

$$i[E^R_x(\boldsymbol{x},t), H^R_y(\boldsymbol{x}', t)] = (1-2\gamma+\dots)\frac{\partial}{\partial z}\delta^3(\boldsymbol{x}-\boldsymbol{x}'). \tag{25.7}$$

It will turn out that the infinities which apparently occur in these expressions are such that the vacuum expectation values of the commutation relations for field intensities averaged over a space-time region are finite,

$$i\left\langle\left[\int_{G_1} F^R_{\mu\varrho}(x')\,\mathrm{d}^4x', \quad \int_{G_2} F^R_{\nu\sigma}(x')\,\mathrm{d}^4x'\right]\right\rangle_0 = \text{finite}. \tag{25.8}$$

The field equations are

$$\left(\gamma^\nu\frac{\partial}{\partial x^\nu}+m\right)\psi = ie\gamma^\nu\psi\Phi_\nu(x),$$

$$\bar{\psi}\left(\gamma^\nu\frac{\overleftarrow{\partial}}{\partial x^\nu}-m\right) = -ie\bar{\psi}\gamma^\nu\Phi_\nu(x),$$

$$\Box\Phi_\nu = -j^\nu = -ie\bar{\psi}\gamma^\nu\psi.$$

Following Yang and Feldman, we make the following decomposition (Eq. [24.1]):

$$\psi(x) = \psi^{\text{in}}(x) - ie\int S^{\text{ret}}(x-x')\,\gamma^\nu\psi(x')\,\Phi_\nu(x')\,\mathrm{d}^4x',$$

$$\bar{\psi}(x) = \bar{\psi}^{\text{in}}(x) - ie\int\bar{\psi}(x')\gamma^\nu S^{\text{adv}}(x'-x)\,\Phi_\nu(x')\,\mathrm{d}^4x',$$

$$\Phi_\nu(x) = \Phi^{\text{in}}_\nu(x) + \int D^{\text{ret}}(x-x')\,j^\nu(x')\,\mathrm{d}^4x'.$$

We wish to calculate $\langle[\Phi_\mu(x), \Phi_\nu(x')]\rangle_0$ up to terms of order e^2. It is sufficient, therefore, in the expansion

$$\Phi_\mu = \Phi_\mu^{(0)} + \Phi_\mu^{(1)} + \Phi_\mu^{(2)} + \ldots$$

in powers of e, to go up to $\Phi_\mu^{(2)}$. We then have

$$[\Phi_\mu(x), \Phi_\nu(x')]^{(2)} = [\Phi_\mu^{(1)}(x), \Phi_\nu^{(1)}(x')] + [\Phi_\mu^{(0)}(x), \Phi_\nu^{(2)}(x\ )] + [\Phi_\mu^{(2)}(x), \Phi_\nu^{(0)}(x')]\ .$$

Because

$$\Phi_\nu^{(2)}(x) = \int D^{\text{ret}}(x-x') j^{\nu(2)}(x')\,\mathrm{d}^4x'\ ,$$

it suffices to calculate $j^{\nu(2)}$. Because of the factor of e in $j^\mu = ie\bar{\psi}\gamma^\mu\psi$, the expansion of j^μ begins with $j^{\mu(1)}$ so that

$$j^{\mu(2)}(x) = ie(\bar{\psi}^{(1)}\gamma^\mu\psi^{(0)} + \bar{\psi}^{(0)}\gamma^\mu\psi^{(1)})\ ,$$

$$\psi^{(1)}(x) = -ie\int S^{\text{ret}}(x-x)\,\gamma^\nu\psi^{(0)}(x')\,\Phi_\nu^{(0)}(x')\,\mathrm{d}^4x'\ .$$

Therefore,

$$j^{\mu(2)}(x) = e^2\int\{\bar{\psi}^{(0)}(x)\gamma^\mu S^{\text{ret}}(x-x')\,\gamma^\nu\psi^{(0)}(x') + \bar{\psi}^{(0)}(x')\gamma^\nu S^{\text{adv}}(x'-x)\,\gamma^\mu\psi^{(0)}(x)\}\,\Phi_\nu^{(0)}(x')\,\mathrm{d}^4x'\ .$$

Now, we have

$$S^{\text{ret}} = \bar{S} - \tfrac{1}{2}S\ ,$$

$$S^{\text{adv}} = \bar{S} + \tfrac{1}{2}S\ .$$

The terms with S contribute nothing to the integral, since there are no real processes in first order. Mathematically,

$$\int S(x-x')\,\gamma^\nu\psi^{(0)}(x')\,\Phi_\nu^{(0)}(x')\,\mathrm{d}^4x' = 0\ ,$$

and

$$\int\bar{\psi}^{(0)}(x')\gamma^\nu S(x'-x)\,\Phi_\nu^{(0)}(x')\,\mathrm{d}^4x' = 0\ .$$

In momentum space, these expressions contain δ-functions (which come from S) which state that both energy and momentum are conserved. However, it is well known that this is not possible for the emission of a photon by a free electron.

Then, upon employing the formula

$$\frac{i}{2}[j^{\mu(1)}(x), j^{\nu(1)}(x')]\varepsilon(x-x') = e^2\{\bar{\psi}^{(0)}(x)\gamma^\mu\bar{S}(x-x')\gamma^\nu\psi^{(0)}(x') + \bar{\psi}^{(0)}(x')\gamma^\nu\bar{S}(x'-x)\gamma^\mu\psi^{(0)}(x)\},$$

we obtain the expression

$$j^{\mu(2)}(x) = \frac{i}{2}\int[j^{\mu(1)}(x), j^{\nu(1)}(x')]\varepsilon(x-x')\Phi_\nu^{(0)}(x')\,\mathrm{d}^4x'. \qquad [25.9]$$

Since

$$\Phi_\mu^{(2)}(x) = \int D^{\mathrm{ret}}(x-x')\,j^{\mu(2)}(x')\,\mathrm{d}^4x',$$

then

$$\Phi_\mu^{(2)}(x) = \frac{i}{2}\int\mathrm{d}^4x'\int\mathrm{d}^4x''\,D^{\mathrm{ret}}(x-x') \cdot[j^{\mu(1)}(x'), j^{\nu(1)}(x'')]\varepsilon(x'-x'')\Phi_\nu^{(0)}(x''), \qquad [25.10]$$

and

$$\Phi_\nu^{(1)}(x) = \int D^{\mathrm{ret}}(x-x')\,j^{\nu(1)}(x')\,\mathrm{d}^4x' = \int\bar{D}(x-x')\,j^{\nu(1)}(x')\,\mathrm{d}^4x',$$

since, from the same considerations as above,

$$\int D(x-x')\,j^{\nu(1)}(x')\,\mathrm{d}^4x' = 0.$$

Now, we wish to go over to the renormalized potential $\Phi_\mu^R(x)$. Then, in accordance with the definition of charge renormalization,

$$\Phi_\mu^{R(2)}(x) = \frac{i}{2}\int\mathrm{d}^4x'\int\mathrm{d}^4x''\,D^{\mathrm{ret}}(x-x')\{[j^{\mu(1)}(x'), j^{\nu(1)}(x'')] - \langle[j^{\mu(1)}(x'), j^{\nu(1)}(x'')]\rangle_0\}\varepsilon(x'-x'')\Phi_\nu^{(0)}(x''). \qquad [25.11]$$

Note: The difficulty with the non-uniquely defined factor of 2 in the charge renormalization is of no consequence, since the idea is that neither $\Phi_\mu^{(2)}$ nor the renormalization, but only the difference, $\Phi_\mu^{R(2)}$, is defined and has significance.

Formally,

$$\frac{i}{2}\int d^4x'\int d^4x'' D^{\text{ret}}(x-x')\langle[j^{\mu(1)}(x'), j^{\nu(1)}(x'')]\rangle_0 \varepsilon(x'-x'')\Phi_\nu^{(0)}(x'')$$
$$= +\int d^4x' D^{\text{ret}}(x-x')\, j^{\mu\,\text{pol}}(x') .$$

(See Sec. 8. There, the field $\mathscr{A}_\mu$ was not quantized. However, formally this makes no difference.)

For the $j^{\mu\,\text{pol}}$, then,

$$j^{\mu\,\text{pol}}(x) = -2\gamma\Box\Phi_\nu^{(0)} + c_1\Box\Box\Phi_\nu^{(0)} + \dots ,$$

where $\gamma+1=e/e_0$. Since $\Phi_\nu^{(0)}$ is a radiation field ($\Box\Phi_\nu^{(0)}=0$), we may certainly drop the higher-order terms. However, the first term is an indeterminate of the form 0/0, as is best seen in momentum space. Because $\Box D^{\text{ret}}(x) = -\delta^4(x)$, formal evaluation using partial integration yields

$$j^{\mu\,\text{pol}}(x) \sim 2\gamma\Phi_\mu^{(0)}(x) .$$

The factor of 2, as stated previously, is not unique. However, this makes no difference in what follows because, as has already been said, $\Phi_\mu^{R(2)}$ is uniquely defined.

We have, therefore,

$$\Phi_\mu^R(x) = \Phi_\mu^{(0)}(x) + \int \overline{D}(x-x')\, j^{\mu(1)}(x')\, d^4x'$$
$$+\frac{i}{2}\int d^4x'\int d^4x'' D^{\text{ret}}(x-x')\big([j^{\mu(1)}(x'), j^{\nu(1)}(x'')]$$
$$-\langle[j^{\mu(1)}(x'), j^{\nu(1)}(x'')]\rangle_0\big)\,\varepsilon(x'-x'')\,\Phi_\nu^{(0)}(x'') .$$

Now, it is the vacuum expectation value of the commutator of this that is of primary interest:

$$\langle [\Phi_\mu^R(x), \Phi_\mu^R(x')] \rangle_0 .$$

It consists of a second-order term in addition to the ones of zeroth-order:

$$i\delta_{\mu\nu} D(x-x') .$$

Note: This is true only in a gauge-invariant expression.

The second-order contribution, which comes from

$$\langle [\Phi_\mu^{R(0)}(x), \Phi_\nu^{R(2)}(x')] \rangle_0 ,$$

vanishes just because of the renormalization. In the vacuum expectation value, the two current commutators cancel directly. There remains only

$$\langle [\Phi_\mu^R(x), \Phi_\nu^R(x')] \rangle_0^{(2)} = \langle [\Phi_\mu^{(1)}(x), \Phi_\nu^{(1)}(x')] \rangle_0 ,$$

or

$$\langle [\Phi_\mu^R(x), \Phi_\nu^R(x')] \rangle_0^{(2)} = \int \mathrm{d}^4x'' \int \mathrm{d}^4x''' \bar{D}(x-x'') \cdot \langle [j^{\mu(1)}(x''), j^{\nu(1)}(x''')] \rangle_0 \bar{D}(x''-x''') . \qquad [25.12]$$

In momentum space, we define

$$i\langle [\Phi_\mu^R(x), \Phi_\nu^R(x')] \rangle_0^{(2)} = \left(\frac{1}{2\pi}\right)^4 \int \exp[ip(x-x')] \Gamma_{\mu\nu}(p) \, \mathrm{d}^4p . \qquad [25.13]$$

Furthermore, in Eq. [7.5] we had

$$\langle [j^{\mu(1)}(x''), j^{\nu(1)}(x''')] \rangle_0 \equiv -2ie^2 K_{\mu\nu}(x''-x''') ,$$
$$= -\frac{2ie^2}{(2\pi)^4} \int \exp[ip(x''-x''')] K_{\mu\nu}(p) \, \mathrm{d}^4p .$$

Since the coordinate-space formula contains only convolutions, the momentum-space relation is simply a product:

$$\Gamma_{\mu\nu}(p) = +\, 2e^2 \frac{1}{(p^2)^2} \cdot K_{\mu\nu}(p) \, . \qquad [25.14]$$

We take $K_{\mu\nu}(p)$ from Eqs. [7.11] and [10.1]:

$$K_{\mu\nu}(p) = i\varepsilon(p)(p_\mu p_\nu - \delta_{\mu\nu} p^2) K(p^2) \, ,$$

$$K(p^2) = \begin{cases} \dfrac{1}{4\pi} \cdot \dfrac{-p^2 + 2m^2}{-3p^2} \sqrt{\dfrac{4m^2 + p^2}{p^2}} \, , & p^2 < -4m^2 \, , \\ 0 \, , & \text{otherwise} \, . \end{cases}$$

The terms proportional to $p_\mu p_\nu$ give no contribution to the gauge-invariant quantities, and are, therefore, actually physically meaningless. We can, for example, consider the field intensities,

$$i\langle [F^R_{\mu\varrho}(x), F^R_{\nu\sigma}(x')] \rangle_0^{(2)} = \left(\frac{1}{2\pi}\right)^4 \int \exp[ip(x - x')] \Gamma_{\mu\varrho\nu\sigma}(p) \, d^4p \, , \qquad [25.15]$$

$$\Gamma_{\mu\varrho\nu\sigma}(p) = p_\mu p_\nu \Gamma_{\varrho\sigma}(p) - p_\nu p_\varrho \Gamma_{\mu\sigma}(p) - p_\mu p_\sigma \Gamma_{\varrho\nu}(p) + p_\varrho p_\sigma \Gamma_{\mu\nu}(p) \, ,$$

and see immediately that the terms proportional to $p_\mu p_\nu$ in $\Gamma_{\mu\nu}(p)$ drop out. There remains, therefore,

$$\Gamma_{\mu\varrho\nu\sigma}(p) = \frac{2e^2}{(p^2)^2} (p_\mu p_\nu \delta_{\varrho\sigma} - p_\nu p_\varrho \delta_{\mu\sigma} - p_\mu p_\sigma \delta_{\varrho\nu} + p_\varrho p_\sigma \delta_{\mu\nu})(-p^2) K(p^2) \, , \qquad [25.16]$$

$$= -\frac{2e^2}{p^2} K(p^2) [p_\mu p_\nu \delta_{\varrho\sigma} - p_\nu p_\varrho \delta_{\mu\sigma} - p_\mu p_\sigma \delta_{\varrho\nu} + p_\varrho p_\sigma \delta_{\mu\nu}] \, . \qquad [25.17]$$

We will continue to calculate with nongauge-invariant quantities, and in doing so we will simply drop $p_\mu p_\nu$ terms.

Relation to the canonical formalism:

$$i\langle [\dot{\Phi}_\mu(\boldsymbol{x}, t), \Phi_\nu(\boldsymbol{x}', t)] \rangle_0^{(2)} = \left(\frac{1}{2\pi}\right)^3 \int \gamma_{\mu\nu}(\boldsymbol{p}) \exp[i\boldsymbol{p} \cdot (\boldsymbol{x} - \boldsymbol{x}')] \, d^3p \, , \qquad [25.18]$$

where

$$\gamma_{\mu\nu}(\boldsymbol{p}) = \frac{1}{2\pi}\int_{-\infty}^{+\infty}(-ip_0)\Gamma_{\mu\nu}(p)\,\mathrm{d}p_0\,. \qquad [25.19]$$

If we drop the $p_\mu p_\nu$ term, then

$$\Gamma_{\mu\nu}(p) = \delta_{\mu\nu}\cdot\Gamma(p)\,, \qquad [25.20]$$

which leads to

$$\gamma_{\mu\nu}(\boldsymbol{p}) = -\frac{i}{2\pi}\delta_{\mu\nu}\int_{-\infty}^{+\infty}p_0\Gamma(p)\,\mathrm{d}p_0\,. \qquad [25.21]$$

Note: In order to be rigorous, we can instead calculate with the field intensities and consider

$$i\langle[E_x(\boldsymbol{x}',t), H_y(\boldsymbol{x}',t)]\rangle_0^{(2)}$$

$$= \left(\frac{1}{2\pi}\right)^3\cdot\frac{-i}{(2\pi)}\int ip_3\,\mathrm{d}^3p\int_{-\infty}^{+\infty}p_0\Gamma(p)\exp[i\boldsymbol{p}\cdot(\boldsymbol{x}-\boldsymbol{x}_0)]\,\mathrm{d}p_0\,. \qquad [25.22]$$

This produces no essential changes in our consideration, for if one compares with the zeroth-order term

$$i\langle[\dot{\Phi}_\mu(\boldsymbol{x},t), \Phi_\nu(\boldsymbol{x}',t)]\rangle_0^{(0)} = \delta_{\mu\nu}\delta^3(\boldsymbol{x}-\boldsymbol{x}')$$

$$= \left(\frac{1}{2\pi}\right)^3\delta_{\mu\nu}\int\exp[i\boldsymbol{p}\cdot(\boldsymbol{x}-\boldsymbol{x}\;)]\,\mathrm{d}^3p$$

and

$$i\langle[E_x(\boldsymbol{x},t), H_y(\boldsymbol{x}',t)]\rangle_0^{(0)} = \frac{\partial}{\partial z}\delta^3(\boldsymbol{x}-\boldsymbol{x}')$$

$$= \left(\frac{1}{2\pi}\right)^3\int ip_3\exp[i\boldsymbol{p}\cdot(\boldsymbol{x}-\boldsymbol{x}')]\,\mathrm{d}^3p$$

then one sees that the second-order term arises from the zeroth-order one by the insertion of the factor

$$\gamma(\boldsymbol{p}) = -\frac{i}{2\pi}\int_{-\infty}^{+\infty}p_0\Gamma(p)\,\mathrm{d}p_0 \qquad [25.23]$$

for the field intensities as well as for the potentials. We will show that this factor is constant:

$$\gamma(\boldsymbol{p}) = \frac{-i}{2\pi}\int_{-\infty}^{+\infty} p_0 \Gamma(p)\,\mathrm{d}p_0$$

$$= \frac{2e^2}{4\pi}\cdot\frac{1}{2\pi}\int_{p^2<-4m^2} p_0\,\varepsilon(p_0)\cdot\frac{1}{-p^2}\cdot\frac{-p^2+2m^2}{-3p^2}\sqrt{\frac{p^2+4m^2}{p^2}}\,\mathrm{d}p_0\,.$$

With

$$p_0\cdot\varepsilon(p_0) = |p_0|$$

and $z = -(p^2+4m^2) > 0$, then,

$$\gamma(\boldsymbol{p}) = \frac{e^2}{4\pi^2}\cdot\frac{1}{3}\int_0^\infty \frac{z+6m^2}{(z+4m^2)^2}\sqrt{\frac{z}{z+4m^2}}\,\mathrm{d}z\,, \qquad [25.24]$$

which is a logarithmically divergent constant. We regularize:

$$\gamma(\boldsymbol{p}) = \gamma(\boldsymbol{p}; m) - \gamma(\boldsymbol{p}; M)\,.$$

Let $z = 4m^2u$ and

$$\gamma_z(\boldsymbol{p}) = \frac{e^2}{4\pi^2}\cdot\frac{1}{3}\int_0^z \mathrm{d}z \ldots\,.$$

Then,

$$\gamma_z(\boldsymbol{p}) = \frac{e^2}{4\pi^2}\cdot\frac{1}{3}\int_0^{z/4m^2} \frac{u+\frac{2}{3}}{(u+1)^2}\sqrt{\frac{u}{u+1}}\,\mathrm{d}u\,, \qquad z \gg 4m^2, 4M^2,$$

$$\tilde{\gamma}(\boldsymbol{p}) = \frac{e^2}{4\pi^2}\cdot\frac{1}{3}\int_{z/4M^2}^{z/4m^2} \frac{u+\frac{2}{3}}{(u+1)^2}\sqrt{\frac{u}{u+1}}\,\mathrm{d}u\,,$$

and with $e^2/4\pi = \alpha$,

$$\tilde{\gamma}(\boldsymbol{p}) \cong \frac{\alpha}{3\pi}\int_{z/4M^2}^{z/4m^2} u^{-1}\mathrm{d}u + O(1) = \frac{\alpha}{3\pi}\log\frac{M^2}{m^2} + O(1)\,,$$

so that

$$\tilde{\gamma}(\boldsymbol{p}) = -2\gamma\,, \qquad [25.25]$$

where

$$\gamma = \frac{\alpha}{3\pi}\log\frac{m}{M}$$

(Dyson-Feynman renormalization). Thus,

$$i\langle[\dot{\Phi}_\mu(\boldsymbol{x},t),\,\Phi_\nu(\boldsymbol{x}',t)]\rangle_0 = \left(\frac{1}{2\pi}\right)^3\delta_{\mu\nu}\int\chi(\boldsymbol{p})\exp[i\boldsymbol{p}\cdot(\boldsymbol{x}-\boldsymbol{x}')]\,\mathrm{d}^3p\,,$$

where

$$\chi(\boldsymbol{p}) = 1 + \gamma(\boldsymbol{p}) + \dots = 1 - 2\gamma + \dots\,.$$

This is a verification, although not physically meaningful. What is essential is that $\Gamma(p^2)$ is finite; that is,

$$i\langle[\Phi^R_\mu(x),\,\Phi^R_\nu(x')]\rangle^{(2)}_0 = \delta_{\mu\nu}\left(\frac{1}{2\pi}\right)^4\int\exp[ip(x-x')]\,\Gamma(p^2)\,\mathrm{d}^4p\,,$$

with finite $\Gamma(p^2)$, while for the nonrenormalized fields there would follow

$$\begin{aligned}&i\langle[\Phi_\mu(x),\,\Phi_\nu(x')]\rangle^{(2)}_0\\ &\quad = \delta_{\mu\nu}\left(\frac{1}{2\pi}\right)^4\int\exp[ip(x-x')][\Gamma(p^2) + 2\gamma\varepsilon(p)\,\delta(p^2+m^2)]\,\mathrm{d}^4p\,,\end{aligned}$$

which is infinite. We obtained a divergent result because the way that the question was put ($t=t'$) was incorrect. Now, on the contrary, if G and G' are four-dimensional volumes (either with sharp boundaries or not), then

$$\begin{aligned}&i\left\langle\left[\int_G\Phi^R_\mu(x)\,\mathrm{d}^4x,\int_{G'}\Phi^R_\nu(x')\,\mathrm{d}^4x'\right]\right\rangle^{(2)}_0\\ &\qquad = \delta_{\mu\nu}\int G(p)\,G(-p)\,\Gamma(p)\,\mathrm{d}^4p\,, \qquad [25.26]\end{aligned}$$

where the factors G make the integral convergent if the volumes are of finite extent in the time coordinate (even for sharp boundaries).

Note: 1. It can reasonably be assumed—although we have not verified it—that analogous relations also hold for spin-zero particles.

2. Dyson has promised a general proof that the same result is valid to any arbitrary degree of approximation [A-7].

3. A somewhat analogous procedure could be carried out for the currents rather than for the fields.[2]

[2] G. KÄLLÉN, *Helv. Phys. Acta* **25**, 417 (1952).

Chapter 6. The S-Matrix: Applications

26. THE RELATION BETWEEN THE S-MATRIX AND THE CROSS-SECTION

In order to apply Dyson's formula to the cross-section, we wrote (Eq. [22.5])

$$U(-\infty, t) = 1 + \int_{-\infty}^{t} W(t')\,\mathrm{d}t' .$$

Using the variables q which commute with H_0, we had written (Eq. [22.6])

$$(q_e|\,W(t)|\,q_a) = \frac{1}{2\pi}\,(q_e|\,R\,|\,q_a)\,\exp\left[i(\omega_e - \omega_a)t\right] ,$$

where a = initial state, e = final state. As mentioned, this is only true *cum grano salis*. The transition probability per unit time then becomes (Eq. [22.10])

$$W = |(q_e|\,R\,|\,q_a)|^2\,\frac{\delta(\omega)}{2\pi} .$$

If we now specialize to the scattering of free particles, then conservation of momentum must hold. This manifests itself in the fact that R contains a δ-function in momentum:

$$(q_e|\,R\,|\,q_a) = (q_e|\,\bar{R}\,|\,q_a)\,\delta^3\Big(\sum_{i=1}^{N'} \boldsymbol{p}_i^a - \sum_{i=1}^{N} \boldsymbol{p}_i^e\Big) , \qquad [26.1]$$

where the N' incoming particles have the momenta $\boldsymbol{p}_1^a$, $\boldsymbol{p}_2^a$, ..., $\boldsymbol{p}_{N'}^a$, while the N outgoing particles have the momenta $\boldsymbol{p}_1^e$, $\boldsymbol{p}_2^e$, ..., $\boldsymbol{p}_N^e$. Here, the particle eigenfunctions are to be thought of as having δ-function normalization:

$$\int \psi_{\boldsymbol{p}}^*(x)\,\psi_{\boldsymbol{p}'}(x)\,\mathrm{d}^3x = \delta^3(\boldsymbol{p}-\boldsymbol{p}') \,. \qquad [26.2]$$

However, in the following it is useful to think of the system as being enclosed in a large box of volume G. The boundary conditions then produce a discrete spectrum, and the normalization is

$$\int_G \Psi_{\boldsymbol{p}}^*(x)\,\Psi_{\boldsymbol{p}'}(x)\,\mathrm{d}^3x = \delta_{\boldsymbol{p},\boldsymbol{p}'} \,, \qquad [26.3]$$

where

$$\delta_{\boldsymbol{p},\boldsymbol{p}'} = \begin{cases} 0 & \boldsymbol{p} \neq \boldsymbol{p}' \,, \\ 1 & \boldsymbol{p} = \boldsymbol{p}' \,. \end{cases}$$

For $\alpha = (2\pi)^3/G$, we obtain $\Psi_{\boldsymbol{p}} = \sqrt{\alpha}\,\psi_{\boldsymbol{p}}$. The matrix element calculated with these functions then becomes

$$(q_e|R_G|q_a) = \alpha^{[(N+N')/2]-1}(q_e|\bar{R}|q_a)\,\delta_{\boldsymbol{P}_a,\boldsymbol{P}_e} \,, \qquad [26.4]$$

where

$$\boldsymbol{P}_a \equiv \sum_1^{N'} \boldsymbol{p}_i^a \,, \qquad \boldsymbol{P}_e \equiv \sum_1^{N} \boldsymbol{p}_i^e \,.$$

Note about the factor $\alpha^{[(N+N')/2]-1}$: A factor of $\sqrt{\alpha}$ appears for every emitted or absorbed particle; the spatial integration which produces $\delta_{\boldsymbol{P}_a,\boldsymbol{P}_e}$ yields the factor α^{-1}.

In the following, we specialize to $N'=2$, the only case in which a cross-section can be meaningfully defined. The transition probability per unit time is

$$\dot{W} = \frac{1}{\pi 2}\cdot \alpha^N \cdot |(q_e|\bar{R}|q_a)|^2\,\delta_{\boldsymbol{P}_a,\boldsymbol{P}_e}\,\delta(\omega) \,. \qquad [26.5]$$

Note: $(\delta_{\boldsymbol{P}_a,\boldsymbol{P}_e})^2 = \delta_{\boldsymbol{P}_a,\boldsymbol{P}_e}$. This was the reason for going over to a finite volume.

The transition probability per unit time, per unit volume, is

$$w = \frac{1}{2\pi}\frac{\alpha^N}{G}|(q_e|\bar{R}|q_a)|^2\delta_{\boldsymbol{P}_a,\boldsymbol{P}_e}\delta(\omega_a - \omega_e)\,. \qquad [26.6]$$

We now require a specification of the initial and final states:

Initial state: $\boldsymbol{p}_1^a$, $\boldsymbol{p}_2^a$ given.

Final state: 1. Here $\mathrm{d}\Omega$ in the direction of $\boldsymbol{p}_1^e$ is given; $|\boldsymbol{p}_1^e|$ is determined (by energy conservation);

$$\mathrm{d}^3p_1^e = (p_1^e)^2\mathrm{d}p_1^e\cdot\mathrm{d}\Omega\,;$$

2. Here $\boldsymbol{p}_2^e$ is completely determined (by momentum conservation);

3. Then $\boldsymbol{p}_3^e, \boldsymbol{p}_4^e, \ldots, \boldsymbol{p}_N^e$ lie in $\mathrm{d}^3p_3^e\mathrm{d}^3p_4^e\ldots\mathrm{d}^3p_N^e$.

For such a process, the probability per unit time, per unit volume, is

$$w = \frac{1}{2\pi}\frac{\alpha^N}{G}\sum|(q_e|\bar{R}|q_a)|^2\delta_{\boldsymbol{P}_a,\boldsymbol{P}_e}\delta(\omega_a - \omega_e)\,, \qquad [26.7]$$

where $\sum$ is to extend over the region in momentum space which corresponds to our final state. Changing this sum back into integrals yields the factors

$$\left.\begin{array}{lll} (1/\alpha)^{N-2} & \text{from} & p_3^e, p_4^e, \ldots, p_N^e \\ 1/\alpha & \text{from} & p_1^e \end{array}\right\},$$

because

$$\sum_{\boldsymbol{p}}\ldots = \frac{1}{\alpha}\int\mathrm{d}^3p\,\ldots\,.$$

Thus,

$$\begin{aligned} w &= \frac{\mathrm{d}\Omega}{2\pi}\frac{(2\pi)^3}{G^2}\int\mathrm{d}p_1^e(p_1^e)^2\delta(\omega_a - \omega_e)|\bar{R}|^2\mathrm{d}^3p_3^e\ldots\mathrm{d}^3p_N^e \\ &= \frac{\mathrm{d}\Omega}{2\pi}\frac{(2\pi)^3}{G^2}\int\mathrm{d}p_1^e\,(p_1^e)^2\delta(\omega_a - \omega_e)|\bar{R}|^2\mathrm{d}^{3(N-2)}p \\ &= \frac{(2\pi)^2}{G^2}\mathrm{d}\Omega\,\mathrm{d}^{3(N-2)}p(p_1^e)^2\left(\frac{\mathrm{d}p_1^e}{\mathrm{d}\omega}\right)|(q_e|\bar{R}|q_a)|^2\,, \end{aligned} \qquad [26.8]$$

where values which satisfy energy and momentum conservation are to be substituted everywhere.

Definition of the cross-section

As a definition of the differential cross-section $\mathrm{d}\sigma$, we write

$$w = \varrho_1^a \varrho_2^a |\boldsymbol{v}_1^a - \boldsymbol{v}_2^a| \cdot \mathrm{d}\sigma \,, \qquad [26.9]$$

where ϱ_1^a, ϱ_2^a = densities of the particles $\boldsymbol{p}_1^a, \boldsymbol{p}_2^a$,

$\boldsymbol{v}_1^a, \boldsymbol{v}_2^a$ = their velocities.

Now, we have

$$\varrho_1^a = \varrho_2^a = \frac{1}{G} \,.$$

Thus,

$$\mathrm{d}\sigma = \frac{(2\pi)^2}{|\boldsymbol{v}_1^a - \boldsymbol{v}_2^a|} \, |(q_e|\bar{R}|q_a)|^2 \cdot (p_1^e)^2 \left(\frac{\mathrm{d}p_1^e}{\mathrm{d}\omega^e}\right) \cdot \mathrm{d}\Omega \, \mathrm{d}^{N-2}p \,. \qquad [26.10]$$

Here,

$$(q_e|S-1|q_a) = (q_e|\bar{R}|q_a)\,\delta(\omega_a - \omega_e)\,\delta^3(\boldsymbol{P}_a - \boldsymbol{P}_e) \,.$$

27. AN APPLICATION OF THE DYSON FORMALISM: MØLLER SCATTERING

Following Dyson, for the Møller scattering of two electrons, we obtain

$$S_2 = \frac{(-i)^2}{2!} \int \mathrm{d}^4x \int \mathrm{d}^4x' \, P(\mathscr{H}_{\text{int}}(x)\,\mathscr{H}_{\text{int}}(x')) \,. \qquad [27.1]$$

From

$$\mathscr{H}_{\text{int}}(x) = -j^\mu \Phi_\mu = -ie(\bar{\psi}\gamma^\mu\psi)\Phi_\mu \,,$$

there follows

$$S_2 = +\frac{e^2}{2} \int \mathrm{d}^4x \int \mathrm{d}^4x' \, P(\bar{\psi}\gamma^\mu\psi \cdot \bar{\psi}'\gamma^\nu\psi') \, P(\Phi_\mu \Phi'_\nu) \,,$$

where

$$\psi \equiv \psi(x) \,, \qquad \psi' \equiv \psi(x') \,, \qquad \text{etc.}$$

The division of P into two parts is allowed because ψ and Φ commute.

We consider the vacuum expectation value of this with respect to the photons:

$$\langle P(\Phi_\mu \Phi'_\nu)\rangle_{0\,\text{ph}} = \tfrac{1}{2}\delta_{\mu\nu} D^C(x - x')\,,$$

which, according to Dyson's theorem, Eq. [15.27], is allowed in such an integral if the kernel satisfies the divergence conditions, as it of course does. We may drop the P which applies to ψ, because we only consider that part which corresponds to the absorption of two electrons and the emission of two others (the "two-electron term," according to Schwinger). Then, we can consider j_μ as commuting, because the commutators involving j_μ are at most "one-electron terms."

For the scattering of two free electrons, we substitute plane waves for ψ:

$$\psi(x) = \left(\frac{1}{2\pi}\right)^{\frac{3}{2}} u(\boldsymbol{p}) \exp[i(px)] a(p)\,,$$

where

$$(px) = \boldsymbol{p}\cdot\boldsymbol{x} - \sqrt{p^2 + m^2}\cdot t\,, \qquad a(p) = \text{absorption operator}\,.$$

We then normalize:

$$\int \psi^*_{\boldsymbol{p}}(x)\,\psi_{\boldsymbol{p}'}(x)\,\mathrm{d}^3x = \delta^3(\boldsymbol{p} - \boldsymbol{p}')\cdot a^*(p)a(p')\,,$$

so that

$$u^*(p)u(p) = 1\,.$$

For ψ, one should substitute the sum of four plane waves which correspond to different momenta:

$$\psi(x) = \psi_{p_1}(x) + \psi_{p_2}(x) + \psi_{p_3}(x) + \psi_{p_4}(x)\,.$$

Since we wish to consider only transitions in which $\boldsymbol{p}_1$ and $\boldsymbol{p}_2$ are absorbed while $\boldsymbol{p}_3$ and $\boldsymbol{p}_4$ are emitted, we can write

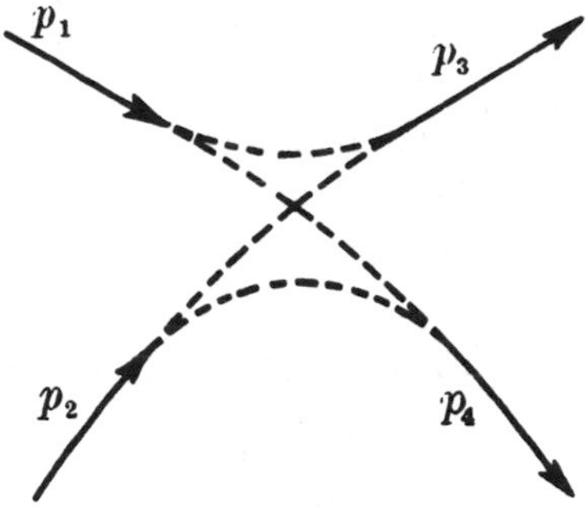

Figure27.1

more simply,

$$\psi(x) = \psi_{p_1} + \psi_{p_2},$$

$$\bar{\psi}(x) = \bar{\psi}_{p_3} + \bar{\psi}_{p_4}.$$

Now, there are four cases:

	ψ	ψ'	$\bar{\psi}$	$\bar{\psi}'$
1	1	2	3	4
2	2	1	3	4
3	1	2	4	3
4	2	1	4	3

Of these, cases 1 and 4, and 2 and 3, respectively, differ only by an interchange of the variables of integration. We can, therefore, restrict ourselves to cases 1 and 2, for example, and put in a factor of 2. Thus

$$S_2 = \frac{e^2}{4} \cdot 2 \int d^4x \int d^4x' \, [(\bar{\psi}_3 \gamma^\mu \psi_1)(\bar{\psi}'_4 \gamma^\mu \psi'_2) + (\bar{\psi}_3 \gamma^\mu \psi_2)(\bar{\psi}'_4 \gamma^\mu \psi'_1)] D^C(x - x').$$

If we substitute the Fourier representation for D^C,

$$D^C(x) = \frac{-2i}{(2\pi)^4} \int \frac{\exp[i(kx)]}{k^2 - i\mu^2} \, d^4k,$$

Eq. [13.12], where $\mu \to 0$ at the end, then we obtain

$$S_2 = \frac{-ie^2}{(2\pi)^2} \int \left[\frac{(\bar{u}_3 \gamma^\mu u_1)(\bar{u}_4 \gamma^\mu u_2)}{k^2 - i\mu^2} \delta^4(p_2 - p_4 + k) \right.$$
$$\left. \cdot \delta^4(p_1 - p_3 - k) - (3 \leftrightarrow 4) \right] d^4k \,.$$

Now, it must be noted that k^2 cannot be zero, as a consequence of the energy-momentum conservation which is expressed by the δ-functions. This is essentially due to the fact that we have calculated with force-free particles (plane waves). Thus, we can perform the integration without considering $-i\mu^2$ to obtain

$$S_2 = \frac{-ie^2}{(2\pi)^2} \left[\frac{(\bar{u}_3 \gamma^\mu u_1)(\bar{u}_4 \gamma^\mu u_2)}{(p_3 - p_1)^2} - \frac{(\bar{u}_3 \gamma^\mu u_2)(\bar{u}_4 \gamma^\mu u_1)}{(p_4 - p_1)^2} \right]$$
$$\cdot \delta^4(p_1 + p_2 - p_3 - p_4) \,. \qquad [27.2]$$

According to our recipe of Eq. [26.10], we obtain

$$d\sigma = \frac{e^4}{(2\pi)^2} \cdot \frac{1}{v} |\boldsymbol{p}_3|^2 \left(\frac{d(\omega_3 + \omega_4)}{d|\boldsymbol{p}_3|} \right)^{-1}$$
$$\cdot \left| \left[\frac{(\bar{u}_3 \gamma^\mu u_1)(\bar{u}_4 \gamma^\mu u_2)}{(p_3 - p_1)^2} - \frac{(\bar{u}_3 \gamma^\mu u_2)(\bar{u}_4 \gamma^\mu u_1)}{(p_4 - p_1)^2} \right] \right|^2 d\Omega \qquad [27.3]$$

for the cross-section.

Note: We have used Heaviside units. Thus, $e^2 = 4\pi/137$.

28. DISCUSSION OF THE D^c FUNCTION [1]

In Møller scattering, the characteristic property of the D^c function was not important because, as a consequence of the absence of forces between the particles, the singularity at $k^2 = 0$ was not at all essential. However, if for ψ we take, for example, the stationary solutions in a static

[1] M. Fierz, *Helv. Phys. Acta* **23**, 731 (1950).

external field, then this peculiarity disappears, and we are in a position to recognize the physically essential properties of the D^C function.

Thus, let

$$\psi(x) = u_n(\boldsymbol{x}) \exp[-i\omega t]\cdot a(\omega) \\ + v_n^*(\boldsymbol{x}) \exp[+i\omega' t]\cdot a^*(\omega') , \qquad \omega, \omega' > 0 .$$

What is essential is that absorption goes with the term $\exp[-i\omega t]$ and emission with $\exp[+i\omega t]$. This is, naturally, also true for $\bar{\psi} = \psi^*\gamma^4$, where

$$\psi^*(x) = u_n^*(\boldsymbol{x}) \exp[+i\omega t]\cdot a^*(\omega) + v_n(\boldsymbol{x}) \exp[-i\omega' t]\cdot a(\omega') ,$$

and is quite generally true for arbitrary fields.

We have now to consider processes in which transitions occur in clearly separated space-time regions V_x and V_y. We want to check whether the following statement is true: If the energy of the charged particles in V_x increases by an amount ω_0, and if the energy in V_y decreases, then V_x is later in time than V_y: $t_x > t_y$. Clearly, this means that

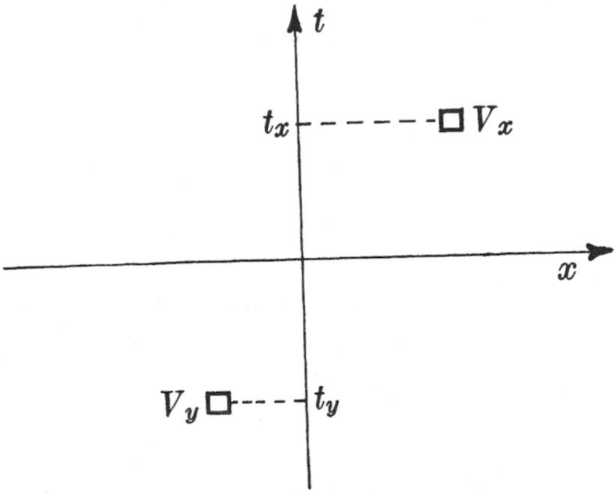

Figure 28.1

photons are emitted before they are absorbed. Now, as a consequence of the uncertainty relation, the uncertainty

$\Delta\omega$ in ω_0, the energy increase in V_x, is greater than $1/T$, where T is the extension of V_x in time:

$$\Delta\omega > \frac{1}{T}.$$

On the other hand, it must be that $\Delta\omega \ll \omega_0$, because otherwise the sign of the energy change is not defined. The statements $\omega_0 > 0$ and $t_x > t_y$ are, therefore, simultaneously meaningful only if

$$T\omega_0 \gg 1. \qquad [28.1]$$

The part of the S-matrix which is due to transitions in V_x and V_y is, according to Section 27,

$$\text{const} \int_{V_x} d^4x \int_{V_y} d^4y \, (\bar{\psi}(x)\,\gamma^\mu \psi(x))\, D^C(x-y)(\bar{\psi}(y)\gamma^\mu\psi(y)).$$

We now consider the processes (1) an increase of the material energy in V_x, and (2) a decrease of the material energy in V_y, and wish to show that 1 must be later than 2, within the restrictions which are guaranteed by the fundamental inequality [28.1]. This is the relationship which the Geneva school calls "causality."[2]

For V_x, we now write

$$\bar{\psi}(x)\,\gamma^\mu\psi(x) \sim a_1 a_2^* \varrho_\mu(\boldsymbol{x}) \exp\left[+i\omega_0 t - \frac{t^2}{T^2}\right];$$

$$\omega_0 \equiv \omega_2 - \omega_1 > 0.$$

Here, $\exp[-t^2/T^2]$ is the function which bounds V_x in the time direction; the spatial part is represented by $\varrho(\boldsymbol{x})$. The central time point of V_x is normalized to $t=0$. The factor $\exp[+i\omega_0 t - t^2/T^2]$ naturally contains negative frequencies too; its Fourier analysis is sketched in Figure 28.2.

Because $\omega_0 T \gg 1$, the curve is very narrow, so that its negative Fourier amplitudes are arbitrarily small. Now,

[2] E. C. G. Stueckelberg and D. Rivier, *Helv. Phys. Acta* **23**, 215 (1950).

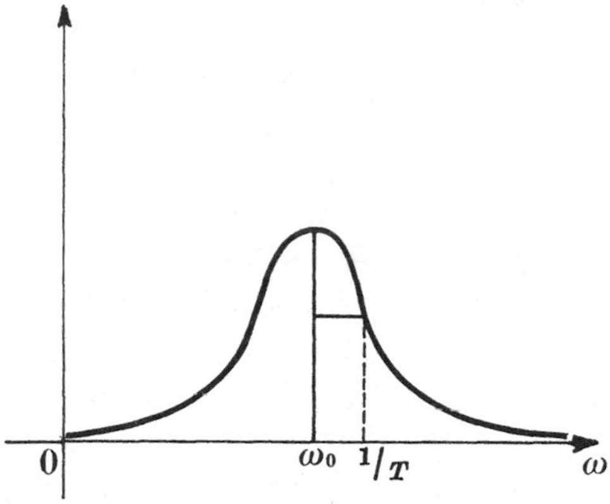

Figure 28.2

we had (compare Eq. [13.15])

$$D^C = -2i[(D^{\text{adv}})^- + (D^{\text{ret}})^+].$$

The $(D^{\text{adv}})^-$ can be readily discarded. It contains only time factors $\exp[+i\omega t]$, so that all together $\exp[+i(\omega_0+\omega)t_x]$ results. These time factors contribute almost nothing to the integral because $T \gg 1/\omega_0$. With $(D^{\text{ret}})^+$ only, our integral becomes

$$\int d^3 x \varrho_\mu(\boldsymbol{x}) \int dt_x \int_{V_y} d^4y \, \frac{1}{8\pi r}\left[\delta(r+t_y-t_x) + \frac{i}{\pi(r+t_y-t_x)}\right] \cdot \exp\left[i\omega_0 \, t_x - \frac{t_x^2}{T^2}\right] \cdot (\bar{\psi}(y)\gamma^\mu \psi(y)),$$

where $r = |\boldsymbol{x}-\boldsymbol{y}|$, and we will show that we can neglect $i/(\pi(r+t_y-t_x))$ as compared with the δ-function term. Then, however, $t_x - t_y = r > 0$, so that $t_x > t_y$. Q.E.D. There then results

$$\int d^3x \, \varrho_\mu(\boldsymbol{x}) \int_{V_y} d^4y \, \frac{1}{4\pi r} \exp\left[i\omega_0(t_y+r) - \left(\frac{t_y+r}{T}\right)^2\right] \cdot (\bar{\psi}(y)\gamma^\mu \psi(y)),$$

and this describes a process at the time $t_y = -r \pm T$; i.e., a "signal on the light cone" having an error of $\pm T$, as demanded by the uncertainty relation.

Furthermore, since V_y is earlier than V_x,

$$|t_y| > T \quad \text{so that} \quad r > T\,,$$

and

$$r\omega_0 \gg 1\,.$$

That is, V_x lies in the wave zone of V_y. *This concept of "causality" has meaning only in the wave zone.*

Now, one can easily show that the neglected term, $i/(\pi(r+t_y-t_x))$, goes as $\sim 1/r^2$ and, therefore, gives $\sim 1/r\omega_0$ times the contribution of the δ-function term. This is the first reason for neglecting this term. In this case, there is a justification for speaking of the free radiation field (real photons). For this reason we needed bound particles, to ensure that energy and momentum conservation do not forbid the emission of real photons.

We can, formally, write this still somewhat differently:

$$D^C = -2i\left(D^{\text{ret}} + (D^{\text{adv}})^- - (D^{\text{ret}})^-\right) = -2i(D^{\text{ret}} + \bar{D}^-)\,.$$

The second term again yields only the time factor

$$\int_{-\infty}^{+\infty} \exp[+i\omega t] \exp\left[+i\omega_0 t - \frac{t^2}{T^2}\right] dt_x\,,$$

which is very small. The first term is exactly the δ-function term considered above. Thus, we have the second justification for neglecting $i/(\pi(r+t_y-t_x))$. It is also seen that the smallness of the above integral, which follows from relation [28.1], and which we have already used above to neglect $(D^{\text{adv}})^-$, is identical with the fact that such terms have no wave zone.

That we have especially considered the relations for photons, and only for the case of Møller scattering, is, naturally, of no essential consequence. What is essential is that D^C, in contrast to its complex conjugate, has the correct "causality property."

29. THE ELECTRON SELF-ENERGY IN AN EXTERNAL HOMOGENEOUS ELECTROMAGNETIC FIELD

As a consequence of the coupling of the electron to the radiation field, the theory, in its present form, yields for the electron a divergent self-energy density of the form

$$\mathscr{H}_s = \delta m \cdot \bar{\psi}(x)\,\psi(x)\,.$$

The idea of mass renormalization, which was originated by Kramers[3] but which was consistently carried out only within the last few years, is that one cannot experimentally separate δm from the mechanical mass m, and thus one must identify $m+\delta m$ with the experimentally measured electron mass. Practically, this means that in all terms which contain $m+\delta m$, the δm is simply to be omitted. The fact that δm is a divergent quantity need not, in principle, be disturbing. The divergence of δm has only the effect that one is no longer certain of the transformation properties with respect to Lorentz transformations; that is, one cannot ascertain whether δm is a scalar or not, so that the prescription for omitting these terms is at first undetermined. The procedure becomes unique only upon including suitable additional specifications (regularization).

This idea of "mass renormalization," together with the analogous "charge renormalization" (see Sec. 12) and suitable regularization prescriptions, suffices to allow reasonable results to be derived from the theory for physically observed quantities. Here, to order e^2, we will calculate the self-energy in an external homogeneous electromagnetic field up to terms which are linear in the field intensities. It will be seen that at first such a divergent, field-independent δm-term appears. However, in addition, there will also be a finite, field-proportional term of the form

$$M = \tfrac{1}{2}\sum_{\mu\nu} F_{\mu\nu}\sigma_{\mu\nu}\,, \qquad [29.1]$$

[3] H. A. Kramers, *Rapports du 8e Conseil Solvay*, 1948 (R. Stoops, Brussels, 1950), p. 241.

where

$$\sigma_{\mu\nu} = \begin{cases} -i\gamma^\mu\gamma^\nu & (\mu \neq \nu) \\ 0 & (\mu = \nu) \end{cases}$$

are the spin operators, and

$$F_{\mu\nu} = \frac{\partial \mathscr{A}_\nu}{\partial x^\mu} - \frac{\partial \mathscr{A}_\mu}{\partial x^\nu}$$

is the external field. That is, there will result

$$\mathscr{H}_S(x) = \big(\bar{\psi}(x)\,\delta m \psi(x)\big) + C\big(\bar{\psi}(x)\,M\psi(x)\big),$$

where

$$C = -\frac{e}{4m}\frac{e^2}{4\pi^2}. \qquad [29.2]$$

The M term represents an observable effect, namely, an addition to the magnetic moment of the electron, which then differs somewhat from the Bohr magneton. There results

$$\mu_{\text{el}} = \left(1 + \frac{\alpha}{2\pi}\right)\mu_B, \qquad [29.3]$$

where

$$\mu_B = \frac{e}{2m} \quad \text{is the Bohr magneton},$$

$$\alpha = \frac{e^2}{4\pi} \simeq \frac{1}{137} \quad \text{is the fine structure constant}.$$

One can also say that the g-factor of the electron is

$$g_{\text{el}} = 2\left(1 + \frac{\alpha}{2\pi}\right). \qquad [29.4]$$

We, therefore, take the field-dependent term in $\mathscr{H}_S(x)$ seriously, while we ascribe no physical significance to the field-independent term $\delta m \bar{\psi}\psi$, since it always appears in the combination $(m + \delta m)$. For us, this means that right from the beginning we have used the incorrect mass.

Note: 1. There exists a proof due to Dyson of the assertion that the renormalization idea works to all orders of perturbation theory.[4]

2. Unfortunately, it has as yet been impossible to carry out the renormalization concept without recourse to perturbation theory [A-8].

Here, we do not follow the original derivation of Schwinger, but, instead, a work of Géhéniau and Villars.[5] This work proceeds from modified Δ and S functions which describe the commutation relations in the case of external fields. (Naturally, the D functions are not changed by the presence of electromagnetic fields.)

The modified functions must satisfy the equations with external fields, as well as the initial conditions without the external fields:

$$\left.\begin{aligned} (\gamma^\mu d_\mu + m)\begin{Bmatrix} S(x,x') \\ S^1(x,x') \end{Bmatrix} &= 0 \\ (\gamma^\mu d_\mu + m)\bar{S}(x,x') &= -\delta^4(x-x') \end{aligned}\right\}, \qquad [29.5]$$

where

$$d_\mu \equiv \frac{\partial}{\partial x^\mu} - ie\mathscr{A}_\mu.$$

Note: The functions are no longer dependent only upon the difference $(x-x')$ of the arguments.

From

$$S^{(\cdots)} = (\gamma^\mu d_\mu - m)\Delta^{(\cdots)},$$

we get

$$\left.\begin{aligned} (d_\mu d_\mu - m^2 + eM)\Delta &= 0 \\ (d_\mu d_\mu - m^2 + eM)\Delta^1 &= 0 \\ (d_\mu d_\mu - m^2 + eM)\bar{\Delta} &= -\delta^4(x-x') \end{aligned}\right\}. \qquad [29.6]$$

In the approximation in which the external field is only

[4] F. J. Dyson, *Phys. Rev.* **75**, 1736 (1949).
[5] J. Géhéniau and F. Villars, *Helv. Phys. Acta* **23**, 179 (1950).

treated linearly, the solution can be written down immediately:

$$\Delta^{(\cdots)}(x, x') = \left[\Delta^{(\cdots)}(x-x') - \frac{\partial \Delta^{(\cdots)}(x-x')}{\partial m^2}\, eM\right] \cdot \exp[ief(x, x')]\,, \qquad [29.7]$$

where for $f(x, x')$ one can give various expressions. Here, we choose a special gauge for the potentials,

$$F_{\mu\nu} = \text{const},$$

so we can choose

$$\mathscr{A}_\mu(x) = -\tfrac{1}{2} F_{\mu\nu} x^\nu . \qquad [29.8]$$

Then, one can write $\xi = x' - x$, and

$$f(x, x') = -\int_x^{x+\xi} \mathscr{A}_\nu(x'')\, \mathrm{d}x''^\nu\,, \qquad [29.9]$$

where the integral is to extend along the straight line

$$x'' = x + \lambda \xi\,.$$

Thus, because of $F_{\mu\nu} = -F_{\nu\mu}$,

$$\left.\begin{aligned} f(x, x') &= -\xi^\nu \int_0^1 \mathscr{A}_\nu(x+\lambda\xi)\,\mathrm{d}\lambda = +\tfrac{1}{2} F_{\nu\varrho}\xi^\nu \int_0^1 (x^\varrho + \lambda \xi^\varrho)\,\mathrm{d}\lambda \\ &= +\tfrac{1}{2} F_{\nu\varrho} x^\varrho \xi^\nu = +\tfrac{1}{2} F_{\nu\varrho} x^\varrho x'^\nu = -\tfrac{1}{2} F_{\nu\varrho} x^\nu x'^\varrho \end{aligned}\right\} . \qquad [29.10]$$

Note: Instead of $\exp[ief]$, in our approximation one could, naturally, have written $(1 + ief)$. However, the exponential form will prove to be favorable for going over to momentum space.

Then, there easily results

$$S^{(\cdots)}(x, x') = (\gamma^\nu d_\nu - m)\Delta^{(\cdots)}(x, x')\,,$$

$$\begin{aligned} S^{(\cdots)}(x, x') = \exp[ief(x, x')] &\left(\gamma^\mu \frac{\partial}{\partial x^\mu} - m\right) \\ &\cdot \left[\Delta^{(\cdots)}(x-x') - \frac{\partial \Delta^{(\ldots)}(x-x')}{\partial m^2}\, eM\right] \\ &+ i\frac{e}{2}\Delta^{(\cdots)}(x-x') F_{\alpha\beta}\gamma^\alpha (x^\beta - x'^\beta)\,. \end{aligned} \qquad [29.11]$$

Our next task is the derivation of a general expression for the self-energy. We effect this with the Dyson formalism. Let $\Phi_\mu(x)$ be the quantized radiation field (in contrast to the external field $\mathscr{A}_\mu(x)$), and let ψ and $\bar{\psi}$ be the electron and positron fields in the presence of the external field. Then, following Dyson,

$$S^{(2)} = \frac{(-i)^2}{2}\int \mathrm{d}^4x \int \mathrm{d}^4x' \, P\{j^\mu(x)j^\nu(x')\Phi_\mu(x)\Phi_\nu(x')\}$$

$$= \frac{(-i)^2}{2}\int \mathrm{d}^4x \int \mathrm{d}^4x' P(j^\mu j'^\nu)P(\Phi_\mu \Phi'_\nu).$$

From this, we take the terms which correspond to the photon vacuum and to the "single-particle terms" with respect to the electrons (that is, those terms for which the average value of the number of particles in the initial and final states is unity):

$$S^{(2)}_{\substack{1\ \mathrm{el}\\ 0\ \mathrm{ph}}} = -\frac{1}{2}\cdot\frac{1}{2}\int \mathrm{d}^4x \int \mathrm{d}^4x' \, \langle P(j^\mu j'^\nu)\rangle_{1\mathrm{el}}\, D^C(x-x'),$$

$$\langle P(j^\mu j'^\nu)\rangle_{1\mathrm{el}} = -e^2 \langle P(\bar{\psi}\gamma^\mu\psi\cdot\bar{\psi}'\gamma^\mu\psi')\rangle_{1\mathrm{el}}\,.$$

In forming the single-particle terms, one has to leave each pair $\bar{\psi}\psi$ with different arguments as it is; the vacuum expectation values of the remaining pairs with different arguments are to be grouped together. (Not grouping together terms with the same argument is a substitute for the neglected subtraction of the vacuum current in the expression $j^\mu = \bar{\psi}\gamma^\mu\psi$.) Thus,

$$\langle P(\bar{\psi}\gamma^\mu\psi\cdot\bar{\psi}'\gamma^\mu\psi')\rangle_{1\mathrm{el}}$$

$$= +\tfrac{1}{2}P(\bar{\psi}'\gamma^\mu\varepsilon(x'-x)\,S^C(x',x)\gamma^\mu\psi + (x \leftrightarrow x')).$$

The P can be dropped if attention is paid to the sign—this exactly cancels the ε. (The terms neglected thereby are no longer of the single-particle type.) Then

$$(-e^2)\,\langle P(\bar{\psi}\gamma^\mu\psi\cdot\bar{\psi}'\gamma^\mu\psi')\rangle_{1\mathrm{el}}$$

$$= +\frac{e^2}{2}P(\bar{\psi}'\gamma^\mu S^C(x',x)\gamma^\mu\psi' + (x\leftrightarrow x)),$$

so that

$$S^{(2)}_{\substack{1\,\mathrm{el}\\0\,\mathrm{ph}}} = -\frac{e^2}{8}\int \mathrm{d}^4x \int \mathrm{d}^4x' \,\{\bar{\psi}(x')\gamma^\mu S^C(x', x)D^C(x'-x)\gamma^\mu\psi(x) + \bar{\psi}(x)\gamma^\mu S^C(x, x')D^C(x-x')\gamma^\mu\psi(x')\},$$

or

$$S^{(2)}_{\substack{1\,\mathrm{el}\\0\,\mathrm{ph}}} = -\frac{e^2}{4}\int \mathrm{d}^4x \int \mathrm{d}^4x' \,(\bar{\psi}(x')\gamma^\mu S^C(x', x) \cdot D^C(x'-x)\gamma^\mu\psi(x)) \,. \qquad [29.12]$$

We now split $S^{(2)}$ into real and imaginary parts. Only the imaginary part is of the self-energy type (see below):

$$\left.\begin{aligned}(S^{(2)}_{\ldots})_I &= +i\frac{e^2}{2}\int \mathrm{d}^4x \int \mathrm{d}^4x' (\bar{\psi}(x)\gamma^\mu[S^1(x, x')\bar{D}(x-x') \\ &\qquad + \bar{S}(x, x')D^1(x-x')]\gamma^\mu\psi(x')) \\ (S^{(2)}_{\ldots})_R &= -\frac{e^2}{2}\int \mathrm{d}^4x \int \mathrm{d}^4x' (\bar{\psi}(x)\gamma^\mu[\tfrac{1}{2}S^1(x, x')D^1(x-x') \\ &\qquad - 2\bar{S}(x, x')\bar{D}(x-x')]\gamma^\mu\psi(x'))\end{aligned}\right\} . \qquad [29.13]$$

Note: 1. The relation between $S_I^{(2)}$ and the self-energy density $\mathscr{H}_S(x)$ is

$$S_I^{(2)} = -i\int \mathscr{H}_S(x)\,\mathrm{d}^4x \,. \qquad [29.14]$$

This is so because if $\mathscr{H}_S$ were in the Hamiltonian, then the contribution to the S-matrix would be exactly the above. However, $\mathscr{H}_S$ is not uniquely determined by this argument. (With respect to this, see remarks 3 and 4 below.)

2. The real part, $S_R^{(2)}$: The unitarity of the S-matrix demands that

$$SS^\dagger = (1 + S^{(1)} + S_I^{(2)} + S_R^{(2)})(1 + S^{(1)\dagger} - S_I^{(2)} + S_R^{(2)}) = 1\,,$$

so that

$$2S_R^{(2)} + S^{(1)}S^{(1)\dagger} = 0\,.$$

Specifically,

$$2\langle S_R^{(2)}\rangle_{\substack{1\,\mathrm{el}\\0\,\mathrm{ph}}} = -\sum_{\mathrm{ph}}\left|\left\langle \begin{matrix}1\,\mathrm{el}\\0\,\mathrm{ph}\end{matrix}\right| S^{(1)} \left|\begin{matrix}1\,\mathrm{el}\\0\,\mathrm{ph}\end{matrix}\right\rangle\right|^2 .$$

Upon integrating over an infinite space-time region, each matrix element of $S^{(1)}$ is to be set equal to zero, since, in this approximation, there are no real processes. An elementary evaluation of the left-hand side also agrees with this statement. However, in order to obviate all ambiguities which originate from questions of convergence, it is useful to perform a regularization (with heavy photons). See p. 148.

3. The imaginary part, $S_I^{(2)}$: It can be written in the following manner:

$$S_I^{(2)} = -i\int \mathscr{H}_S(x)\,\mathrm{d}^4x\,,$$

$$\mathscr{H}_S(x) = -\frac{e^2}{4}\int \mathrm{d}^4x'\,\{\bar{\psi}(x')\gamma^\mu[S^{(1)}(x',x)\,\bar{D}(x'-x) + \bar{S}(x',x)\,D^1(x'-x)]$$
$$\cdot\gamma^\mu\psi(x) + \text{h.c.}\}\,.$$

Above, we have already brought up the question of to what extent this is a self-energy density. Anyhow, the self-energy density is meaningless, so that only the total energy,

$$H_S = \int \mathscr{H}_S(x)\,\mathrm{d}^3x\,,$$

interests us. However, with $S_I^{(2)}$ we obtain even less, namely, only

$$\int H_S\,\mathrm{d}t = +iS_I^{(2)}\,.$$

This is characteristic of the Dyson formalism, which is primarily convenient for scattering processes. In its favor is to be said that terms in H_S whose time integrals vanish are, anyhow, not to be considered as self-energies but should be viewed as fluctuations (*Zitterbewegungen*), so that they may be omitted. (This fluctuation is oscillatory, $\sim\exp[2i\omega t]$, and corresponds to the creation of virtual pairs.) Furthermore, in the above form H_S is so chosen that it contains no *Zitterbewegung* for a force-free electron. However, in our case of external fields, there are such terms, and we will drop them.

4. It may be of interest to derive the self-energy using still another type of formalism (Schwinger). This derivation gives us H_S, but, however, not the real part of $S^{(2)}$. In the interaction representation,

$$i\frac{\partial\Psi}{\partial t} = H_{\text{int}}\,\Psi, \qquad \mathscr{H}_{\text{int}} = -j^\nu\,\Phi_\nu = -ie(\bar{\psi}\gamma^\nu\psi)\,\Phi_\nu\,.$$

We would like to remove the H_{int} by means of a canonical transformation:

$$\Psi = e^S \Psi' .$$

If one chooses $i\dot{S} = H_{\text{int}}$, then

$$i \frac{\partial \Psi'}{\partial t} = \frac{1}{2} [H_{\text{int}}, S] \Psi' .$$

This is easily obtained from the Lie series

$$e^{-S} O e^{+S} = O + [O, S] + \frac{1}{2!} [[O, S], S] + \dots$$

$$= \sum_{n=0}^{\infty} \frac{1}{n!} [[\dots[[O, \underbrace{S], S], \dots, S], S}_{n}] . \qquad [29.15]$$

Proof:

$$e^{-S} O e^{+S} = \sum_{nm} \frac{(-1)^n}{n!m!} S^n O S^m = \sum_{N=0}^{\infty} \frac{1}{N!} \sum_{n=0}^{N} (-1)^n \binom{N}{n} S^n O S^{N-n} .$$

By induction, on the other hand, one easily finds that

$$[\dots[[O, \underbrace{S], S], \dots, S]}_{n} = \sum_{\nu=0}^{m} (-1)^\nu \binom{n}{\nu} S^\nu O S^{n-\nu} .$$

Thus,

$$e^{-S} O e^{S} = \sum_{N=0}^{\infty} \frac{1}{N!} [\dots[[O, \underbrace{S], S], \dots, S]}_{N} . \qquad \text{Q.E.D.}$$

Then, for $O = H_{\text{int}}$,

$$e^{-S} H_{\text{int}} e^S = H_{\text{int}} + [H_{\text{int}}, S] \dots ,$$

while for $O = \partial/\partial t$,

$$e^{-S} \frac{\partial}{\partial t} e^S = \frac{\partial}{\partial t} + \frac{\partial S}{\partial t} + \frac{1}{2} \left[\frac{\partial S}{\partial t}, S \right] + \dots .$$

Then, with $\Psi = e^S \Psi'$,

$$i \frac{\partial \Psi}{\partial t} = H_{\text{int}} \Psi$$

becomes

$$i \left(e^{-S} \frac{\partial}{\partial t} e^S \right) \Psi' = (e^{-S} H_{\text{int}} e^S) \Psi' .$$

Upon substituting the series, the result is obtained. From $\dot{S} = -iH_{\text{int}}$, there follows

$$S(t) = -i\int_{-\infty}^{t} H_{\text{int}}(t')\,\mathrm{d}t',$$

or

$$S(t) = -\frac{i}{2}\int_{-\infty}^{+\infty}(1+\varepsilon(t-t'))\,H_{\text{int}}(t')\,\mathrm{d}t'.$$

Now, since H_{int} contains no real processes of first order,

$$\int_{-\infty}^{+\infty} H_{\text{int}}(t')\,\mathrm{d}t' = 0\,,$$

it follows that

$$S(t) = -\frac{i}{2}\int_{-\infty}^{+\infty}\varepsilon(t-t')\,H_{\text{int}}(t')\,\mathrm{d}t'. \qquad [29.16]$$

Thus,

$$\frac{1}{2}[H_{\text{int}}, S] = -\frac{i}{4}\int_{-\infty}^{+\infty}\varepsilon(t-t')[H_{\text{int}}(t), H_{\text{int}}(t')]\,\mathrm{d}t',$$

$$\frac{1}{2}[H_{\text{int}}, S] = +\frac{ie^2}{4}\int \mathrm{d}^3x\int \mathrm{d}^4x'\,\varepsilon(x-x')[\bar{\psi}\gamma^\mu\psi\Phi_\mu, \bar{\psi}'\gamma^\nu\psi'\Phi'_\nu]\,.$$

Now, the following auxiliary formulas are useful. If

$$[A, a] = [A, b] = [B, a] = [B, b] = 0\,,$$

or if

$$\{A, a\} = \{A, b\} = \{B, a\} = \{B, b\} = 0\,,$$

that is, if A and B either commute *or* anticommute with a and b, then

$$\left.\begin{aligned}[Aa, bB] &= \tfrac{1}{2}[A, B]\{a, b\} + \tfrac{1}{2}\{A, B\}[a, b]\\ \{Aa, bB\} &= \tfrac{1}{2}[A, B][a, b] + \tfrac{1}{2}\{A, B\}\{a, b\}\end{aligned}\right\}. \qquad [29.17]$$

Then

$$[\bar{\psi}\gamma^\mu\psi\Phi_\mu, \bar{\psi}'\gamma^\nu\psi\Phi'_\nu]$$
$$= \tfrac{1}{2}[\bar{\psi}\gamma^\mu\psi, \bar{\psi}'\gamma^\nu\psi']\{\Phi_\mu, \Phi'_\nu\} + \tfrac{1}{2}\{\bar{\psi}\gamma^\mu\psi, \bar{\psi}'\gamma^\nu\psi'\}[\Phi_\mu, \Phi'_\nu]\,.$$

If we consider the zero-photon term, then we get

$$\tfrac{1}{2}[\bar{\psi}\gamma^\mu\psi, \bar{\psi}'\gamma^\mu\psi']D^1(x-x') + \frac{i}{2}\{\bar{\psi}\gamma^\mu\psi, \bar{\psi}'\gamma^\mu\psi'\}D(x-x')\,.$$

From this, one has now to construct the one-electron term, which, upon using Eq. [29.17], proceeds in a manner completely analogous to that in the Dyson formalism. One then obtains

$$\langle[\bar{\psi}\gamma^\mu\psi, \bar{\psi}'\gamma^\mu\psi']\rangle_{1\,\mathrm{el}} = -i(\bar{\psi}\,\gamma^\mu S(x, x')\gamma^\mu\psi' - (x \leftrightarrow x')) ,$$

$$\langle\{\bar{\psi}\gamma^\mu\psi, \bar{\psi}'\gamma^\mu\psi'\}\rangle_{1\,\mathrm{el}} = -(\bar{\psi}\gamma^\mu S^1(x, x')\gamma^\mu\psi' + (x \leftrightarrow x')) .$$

Then,

$$\tfrac{1}{2}\langle[H_{\mathrm{int}}, S]\rangle_{\substack{0\,\mathrm{ph}\\ 1\,\mathrm{el}}} = +\frac{e^2}{8}\int \mathrm{d}^3x\int \mathrm{d}^4x'\,\varepsilon(x-x')(\bar{\psi}\gamma^\mu[S(x, x')D^1(x-x')$$

$$+ S^1(x, x')D(x-x')]\gamma^\mu\psi' - (x \leftrightarrow x'))$$

$$= -\frac{e^2}{4}\int \mathrm{d}^3x\int \mathrm{d}^4x'\{\bar{\psi}\gamma^\mu[\bar{S}D^1 + S^1\bar{D}]\gamma^\mu\psi' + (x \leftrightarrow x')\} .$$

If we now note that

$$H_S = \tfrac{1}{2}\langle[H_{\mathrm{int}}, S]\rangle_{\substack{0\,\mathrm{ph}\\ 1\,\mathrm{el}}} , \qquad [29.18]$$

by definition, then we obtain exactly the same expression as we did previously with the Dyson formalism.

a. Evaluation of the self-energy

In momentum space,

$$\begin{cases} \bar{\Delta}\,(k) = \dfrac{1}{k^2 + m^2} \\ \Delta^1(k) = 2\pi\,\delta(k^2 + m^2) , \end{cases}$$

where one always is to take the principal value with respect to the k_0 integration. In the force-free case, the following lemma is useful:

$$\int \exp[-i(q\xi)][\bar{D}(\xi)\Delta^1(\xi) + D^1(\xi)\bar{\Delta}(\xi)]\,\mathrm{d}^4\xi$$

$$= \left(\frac{1}{2\pi}\right)^4\int [\bar{D}(k)\Delta^1(q-k) + D^1(k)\bar{\Delta}(q-k)]\,\mathrm{d}^4k$$

$$= \left(\frac{1}{2\pi}\right)^3\int \left[\frac{\delta((q-k)^2 + m^2)}{k^2} + \frac{\delta(k^2)}{(q-k)^2 + m^2}\right]\mathrm{d}^4k ,$$

and, because of the properties of the δ-function,

$$\left[\frac{\delta\big((q-k)^2+m^2\big)}{k^2}+\frac{\delta(k^2)}{(q-k)^2+m^2}\right]$$

$$=-\frac{\delta\big((q-k)^2+m^2\big)}{(q^2+m^2-2kq)}+\frac{\delta(k^2)}{(q^2+m^2-2kq)}$$

$$=-\int_0^1\delta'[k^2+v(q^2+m^2-2kq)]\,\mathrm{d}v\,,$$

so that

$$\int\exp[-i(q\xi)][\bar{D}(\xi)\Delta^1(\xi)+D^1(\xi)\bar{\Delta}(\xi)]\,\mathrm{d}^4\xi$$

$$=-\left(\frac{1}{2\pi}\right)^3\int\mathrm{d}^4k\int_0^1\delta'\big(k^2+v(q^2+m^2-2kq)\big)\,\mathrm{d}v\,. \qquad [29.19]$$

Remark: Something very analogous can be done with the formula for $S^{(2)}$ which contains the real and imaginary parts:

$$\frac{1}{2}\int\exp[-i(q\xi)]D^C(\xi)\,\Delta^C(\xi)\,\mathrm{d}^4\xi=\frac{1}{2}\left(\frac{1}{2\pi}\right)^4\int D^C(k)\,\Delta^C(q-k)\,\mathrm{d}^4k$$

$$=\frac{-2}{(2\pi)^4}\int\frac{\mathrm{d}^4k}{(k^2-i\mu^2)\big((q-k)^2+m^2-i\mu^2\big)}\,,$$

where, thanks to the μ, the k_0 integration is also to extend over the real axis.

We can transform this with Feynman's relation, which was already used in Eq. [11.2],

$$\frac{1}{ab}=\int_0^1\frac{\mathrm{d}v}{[b+(a-b)v]^2}=\int_0^1\frac{\mathrm{d}v}{[a+(b-a)v]^2}\,,$$

so that

$$\frac{1}{2}\int\exp[-i(q\xi)]D^C(\xi)\,\Delta^C(\xi)\,\mathrm{d}^4\xi$$

$$=\frac{-2}{(2\pi)^4}\int\mathrm{d}^4k\int_0^1\frac{\mathrm{d}v}{[k^2+i\mu^2+v(q^2+m^2-2kq)]^2}\,. \qquad [29.20]$$

In the same way, we can evaluate $S^{(2)}$ for the force-free case. From

$$\psi(x) = \int u(q) \exp[-i(qx)]\, d^4q$$

and $\xi = x' - x$, we have (see Eq. [29.13])

$$S^{(2)} = -\frac{e^2}{2} \int \bar{u}(q) \exp[-i(qx)] F(q)\, \psi(x)\, d^4q\,, \qquad [29.21]$$

where

$$F(q) = \frac{1}{2} \int \exp[-i(q\xi)] \gamma^\mu S^C(\xi)\, \gamma^\mu D^C(\xi)\, d^4\xi$$

$$= \frac{-2}{(2\pi)^4} \int \frac{\gamma^\mu [i\gamma(q-k) - m] \gamma^\mu}{(k^2 - i\mu^2)\big((q-k)^2 + m^2 - i\mu^2\big)} d^4k\,.$$

From

$$\gamma^\mu [i\gamma(q-k) - m] \gamma^\mu = -2i\gamma(q-k) - 4m$$

and Eq. [11.2], we conclude that

$$F(q) = \frac{-2}{(2\pi)^4} \int d^4k \int_0^1 \frac{-2i\gamma(q-k) - 4m}{[k^2 - i\mu^2 + v(q^2 + m^2 - 2kq)]^2} dv\,. \qquad [29.22]$$

Note: These are for the moment only conditionally correct formulas, since the k integration diverges. However, when one regularizes suitably, the k integration becomes convergent.

The "regularization"[6] is a formal procedure for handling divergent integrals without losing Lorentz invariance. In the simplest case of this procedure, in addition to real photons, one has also to couple photons of large mass M using an imaginary coupling constant ie. This yields

$$\frac{1}{k^2 - i\mu^2} \to \frac{1}{k^2 - i\mu^2} - \frac{1}{k^2 + M^2 - i\mu^2}\,,$$

[6] R. P. Feynman, *Phys. Rev.* **76**, 769 (1949); W. Pauli and F. Villars, *Rev. Mod. Phys.* **21**, 434 (1949).

and, in the integrand of Eq. [29.22],

$$\frac{1}{[\ldots]^2} \to \left(\frac{1}{[k^2 - i\mu^2 + v(q^2 + m^2 - 2kq)]^2} - \frac{1}{[k^2 + M^2 - i\mu^2 + v(q^2 + m^2 - 2kq - M^2)]^2} \right).$$

With this our integral becomes

$$\frac{-2}{(2\pi)^4} \int d^4k \int_0^1 \left(\frac{-2i\gamma(q-k) - 4m}{[k^2 - i\mu^2 + v(q^2 + m^2 - 2kq)]^2} - \frac{-2i\gamma(q+k) - 4m}{[k^2 + M^2 - i\mu^2 + v(q^2 + m^2 - 2kq - M^2)]^2} \right) dv\,,$$

which converges. We can evaluate this convergent integral in the complex plane by closing the path in the upper half-plane. The contribution over the semicircle vanishes,

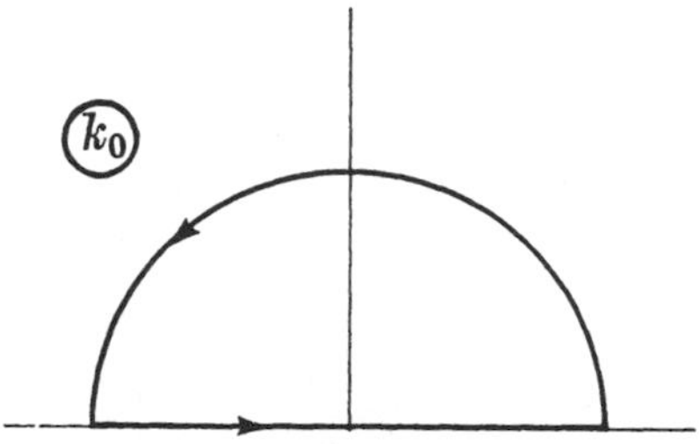

Figure 29.1

while the sum of the residues yields an imaginary contribution which agrees exactly with Schwinger's expression obtained above. This is a proof for the previous statement that the real part of $S^{(2)}$ vanishes upon regularization.

Now. back to the calculation of Villars and Géhéniau.

With $\lambda \equiv (x^\alpha - x'^\alpha)(x^\alpha - x'^\alpha)$, we had (Eqs. [29.11], [29.9])

$$\bar{S}(x, x') = \exp\left[ie\int_{x'}^{x} \mathscr{A}''_\nu \mathrm{d}x''^\nu\right]\left(\gamma\frac{\partial}{\partial x} - m\right)\left[\bar{\Delta}(\lambda) - \frac{\partial\bar{\Delta}(\lambda)}{\partial m^2} eM\right] + \frac{ie}{2}\bar{\Delta}(\lambda) F_{\alpha\beta}\gamma^\alpha(x^\beta - x'^\beta),$$

and an analogous expression for S^1. The integral in the exponent is to extend along the straight line. Then,

$$\mathscr{H}_S(x) = -\frac{e^2}{4}\int \mathrm{d}^4\xi\,\{\bar{\psi}(x+\xi)\gamma^\mu[\bar{D}(\lambda)S^1(x+\xi, x) + D^1(\lambda)\bar{S}(x+\xi, x)]\,\gamma^\mu\psi(x) + \text{h.c.}\}.$$

Note: No vacuum polarization will appear, since the external field has no sources.

Upon substituting the expressions for S^1 and S, one obtains

$$\mathscr{H}_S(x) = I + II + III.$$

Here,

$$I = -\frac{e^2}{4}\int \mathrm{d}^4\xi\left\{\bar{\psi}(x+\xi)\gamma^\mu[\bar{D}(\lambda)\Gamma(\xi)\Delta^1(\lambda) + D^1(\lambda)\Gamma(\xi)\bar{\Delta}(\lambda)]\,\gamma^\mu\psi(x)\cdot\exp\left[ie\int_{x'}^{x}\mathscr{A}''_\nu \mathrm{d}x''\right] + \text{h.c.}\right\}, \quad [29.23]$$

where

$$\Gamma(\xi) \equiv \left(\gamma^\mu\frac{\partial}{\partial\xi^\mu} - m\right); \quad [29.24]$$

$$II = +\frac{e^3}{4}\int \mathrm{d}^4\xi\left\{\bar{\psi}(x+\xi)\,\gamma^\mu\left[\bar{D}(\lambda)\Gamma(\xi)\,\frac{\partial\Delta^1}{\partial m^2} + D^1(\lambda)\Gamma(\xi)\,\frac{\partial\bar{\Delta}}{\partial m^2}\right]\cdot M\gamma^\mu\psi(x) + \text{h.c.}\right\}; \quad [29.25]$$

and

$$III = +\frac{ie^3}{8}F_{\mu\nu}\int \mathrm{d}^4\xi\,\{\bar{\psi}(x+\xi)\,\gamma^\alpha[\bar{D}(\lambda)\,\Delta^1(\lambda) + D^1(\lambda)\,\bar{\Delta}(\lambda)]\cdot\gamma^\mu\gamma^\alpha\xi^\nu\psi(x) + \text{h.c.}\}. \quad [29.26]$$

Note: Here, we have neglected terms of order higher than e^4, except in the exponential of I. There, it turns out to be practical to keep them.

b. Evaluation of I

We write this term partially in momentum space, leaving, however, the $\psi(x)$ as it is:

$$I = \frac{e^2}{2\cdot(2\pi)^4}\int d^4q\, \overline{u}(q) \exp[-i(qx)] \int d^4k \left\{[\overline{D}(k)\Delta^1(k-\tilde{q}) \right.$$
$$\left. + D^1(k)\overline{\Delta}(k-\tilde{q})]\cdot\gamma^\alpha[+i\gamma(k-\tilde{q})+m]\gamma^\alpha\psi(x) + \text{c.c.}\right\},$$

where c.c. represents the charge-conjugated expression. Here,

$$\tilde{q} = q - e\mathscr{A}(x) .$$

If

$$d_\mu = \frac{\partial}{\partial x^\mu} + ie\mathscr{A}_\mu(x) ,$$

then

$$-d_\mu\overline{\psi}\gamma^\mu + m\overline{\psi} = 0 .$$

That is,

$$\int \exp[-i(qx)]\overline{u}(q)(i\gamma\tilde{q}+m)\, d^4q = 0 .$$

Now, we can use our lemma of Eq. [29.19], and introduce the parameter v. The δ' is then

$$\delta'[k^2 + v(\tilde{q}^2 + m^2 - 2k\tilde{q})] .$$

One now makes a translation of the variables of integration,

$$k' = k - v\tilde{q} , \qquad [29.27]$$

so that

$$\delta'(k'^2 + v^2m^2 + \varepsilon) ,$$

with

$$\varepsilon = (v - v^2)(\tilde{q}^2 + m^2)$$

results.

Note: This type of translation (described by Eq. [29.27]) is permitted only for sufficiently convergent integrals (that is, those with less than a logarithmic divergence), and, therefore, is not really permitted here. However, since the field-dependent term converges, as we shall see, it is proper in this case; on the contrary, the actual field-independent self-energy is definable in a Lorentz-invariant way only after regularization.

Furthermore,

$$\gamma^{\alpha}[+i\gamma(k-\tilde{q})+m]\gamma^{\alpha}=-2i\gamma(k-\tilde{q})+4m$$
$$=-2i\gamma\big(k'-\tilde{q}(1-v)\big)+4m\,.$$

The term which is linear in k' will, from symmetry arguments, contribute nothing, so that finally

$$I=-\frac{e^2}{2\cdot(2\pi)^3}\int \mathrm{d}^4k\int_0^1 \mathrm{d}v\int \mathrm{d}^4q\{\overline{u}(q)\exp[-i(qx)]\delta'(k^2+m^2v^2+\varepsilon)$$
$$\cdot[-i\gamma\tilde{q}(v-1)+2m]\psi(x)+\text{h.c}\}\,. \qquad [29.28]$$

Consider the part that goes as e^2 (not containing the external field):

$$\tilde{q}\to q\,,\qquad q^2+m^2=0;\qquad \text{hence}\qquad \varepsilon=0\,,\qquad i\gamma q=-m\,,$$

so that

$$I^0=-\frac{e^2}{2\cdot(2\pi)^3}\int \mathrm{d}^4k\int_0^1 \mathrm{d}v\int \mathrm{d}^4q$$
$$\cdot\overline{u}(q)\exp[-i(qx)]\delta'(k^2+m^2v^2)\,m(v+1)\,\psi(x)\,.$$

The integral

$$I^0\equiv\int \overline{u}(q)\exp[-i(qx)]I^0_q\cdot\psi(x)\,\mathrm{d}^4q$$

gives the divergent self-energy of the electron, which, as already mentioned, can be rigorously defined only by regularization.

We have

$$\int \delta'(-k_0^2+A)\,\mathrm{d}k_0=\int_0^{\infty}\delta'(-z+A)\frac{\mathrm{d}z}{\sqrt{z}}=-\frac{1}{2}A^{-\frac{3}{2}},$$

so that

$$\int \delta'(k^2 + m^2v^2)\, \mathrm{d}^4k$$

$$= -\frac{1}{2}\int \frac{\mathrm{d}^3k}{(k^2 + m^2v^2)^{\frac{3}{2}}} = -\, 2\pi \int_0^\infty \frac{k^2}{(k^2 + m^2v^2)^{\frac{3}{2}}}\, \mathrm{d}k\ .$$

Formally, with $k = mvz$, the integral above becomes

$$-\, 2\pi \int_0^\infty \frac{z^2}{(z^2 + 1)^{\frac{3}{2}}}\, \mathrm{d}z\ ,$$

which is logarithmically divergent. Then

$$I_q^0 = \frac{e^2}{2\cdot(2\pi)^2}\cdot\frac{3}{2}\, m \int_0^\infty \frac{z^2}{(1 + z^2)^{\frac{3}{2}}}\, \mathrm{d}z\ , \qquad [29.29]$$

which, of course, is undefined.

Consider the part which goes as e^3 (linear terms in the external field). We write $I = I^0 + I^1 + \text{c.c.}$, where I^1 then goes as e^3. One must expand in powers of ε up to linear terms:

$$I^1 = -\frac{e^2}{2\cdot(2\pi)^3}\int \bar{u}(q)\exp[-i(qx)]I_q^1\cdot\psi(x)\, \mathrm{d}^4q\ ,$$

where

$$I_q^1 = \int_0^1 \mathrm{d}v \int \mathrm{d}^4k\, \delta''(k^2 + m^2v^2)\, [i\gamma\tilde{q}(1 - v) + 2m]v(1 - v)(\tilde{q}^2 + m^2)\ .$$

Now,

$$\int \delta''(-k_0^2 + A)\, \mathrm{d}k_0 = \int_0^\infty \delta''(z + A)\, \frac{\mathrm{d}z}{\sqrt{z}} = +\frac{1}{2}\cdot\frac{3}{2}\, A^{-\frac{5}{2}}\ ,$$

so that

$$\int \delta''(k^2 + m^2v^2)\, \mathrm{d}^4k = \frac{3}{4}\cdot 4\pi \int_0^\infty \frac{k^2\mathrm{d}k}{(k^2 + m^2v^2)^{\frac{5}{2}}} = 3\pi\, \frac{1}{3v^2m^2} = \frac{\pi}{v^2m^2}\ ,$$

since

$$\int_0^\infty \frac{k^2}{(k^2+a^2)^{\frac{5}{2}}}\,\mathrm{d}k = \frac{1}{3}\int_0^\infty \frac{\mathrm{d}k}{(k^2+a^2)^{\frac{3}{2}}} = \frac{1}{3a^2}\,.$$

Thus,

$$I_q^1 = \frac{\pi}{m^2}\int_0^1 [i\gamma\tilde{q}(1-v)+2m]\,\frac{1-v}{v}\,(\tilde{q}^2+m^2)\,\mathrm{d}v\,. \qquad [29.30]$$

Note: The integrand has a singularity at $v=0$ which will cancel with other terms.

Now, we decompose further:

$$I_q^1 = I_q^{1'} + I_q^{1''}, \qquad [29.31]$$

where

$$I_q^{1'} = -\frac{e^2}{4\cdot(2\pi)^2}\cdot\frac{1}{m}(\tilde{q}^2+m^2)\int_0^1 \frac{1-v^2}{v}\,\mathrm{d}v\,,$$

$$I_q^{1''} = -\frac{e^2}{4\cdot(2\pi)^2}\,\frac{1}{m^2}\,(i\gamma\tilde{q}+m)(\tilde{q}^2+m^2)\int_0^1 \frac{(1-v)^2}{v}\,\mathrm{d}v\,.$$

Because of the Dirac equation,

$$\int \overline{u}(q)\exp[-i(qx)](\tilde{q}^2+m^2-eM)\,\mathrm{d}^4q = 0\,.$$

Thus, in $I_q^{1'}$ we can make the substitution

$$\tilde{q}^2+m^2 = eM\,.$$

We obtain

$$I_q^{1'} = -\frac{e}{4m}\,\frac{e^2}{4\pi^2}\,M\int_0^1 \frac{1-v^2}{v}\,\mathrm{d}v\,.$$

Thus,

$$(I_q^{1'} + \text{c.c.}) = -\frac{e}{2m}\,\frac{e^2}{4\pi^2}\,(\bar{\psi}M\psi)\int_0^1 (1-v^2)\,\frac{\mathrm{d}v}{v}\,. \qquad [29.32]$$

The term $I_q^{1''}$ is not of the self-energy type; instead it describes a *Zitterbewegung*. From

$$\int \overline{u}(q)(i\gamma\tilde{q}+m)\exp[-iqx]\,\mathrm{d}^4q = 0\,,$$

upon application of

$$(-d_\mu d_\mu + m^2)\,,$$

there follows

$$0=\int \exp[-i(qx)]\overline{u}(q)\,[(\tilde{q}^2+m^2)(i\gamma\tilde{q}+m)-eF_{\alpha\beta}\gamma^\alpha\tilde{q}_\beta]\,\mathrm{d}^4q\,.$$

(Note that the $\mathscr{A}(x)$ contained in $\tilde{q}$ must also be differentiated.) With this,

$$I_q^{1''}=-\frac{e}{4m^2}\frac{e^2}{4\pi^2}F_{\alpha\beta}\gamma^\alpha\tilde{q}_\beta\int_0^1\frac{(1-v)^2}{v}\,\mathrm{d}v\,.$$

Later, we will find more such terms and show that they correspond to a *Zitterbewegung*.

c. *Evaluation of II*

We had

$$II=\frac{e^3}{4}\left(\frac{1}{2\pi}\right)^4\int \mathrm{d}^4k\int \mathrm{d}^4q\,\Bigg\{\overline{u}(q)\exp[-i(qx)]$$
$$\cdot\left[\overline{D}(k)\,\frac{\partial\Delta^1(k-q)}{\partial m^2}+D^1(k)\,\frac{\partial\overline{\Delta}(k-q)}{\partial m^2}\right]$$
$$\cdot\gamma^\alpha[i\gamma(q-k)-m]M\gamma^\alpha\psi(x)+\text{h.c.}\Bigg\}\,. \qquad [29.33]$$

Here, because of the e^3, we can always calculate with the force-free Dirac equation. We have

$$M\gamma^\alpha=\gamma^\alpha M+2iF_{\alpha\beta}\gamma^\beta\,,$$
$$\gamma^\alpha(i\gamma p-m)\gamma^\alpha=-2(i\gamma p+2m)\,,$$
$$-2iF_{\alpha\beta}\gamma^\alpha\gamma^\beta=4M\,,$$

so that

$$\gamma^\alpha[i\gamma(q-k)-m]M\gamma^\alpha = -2[i\gamma(q-k)+2m]M$$
$$+\,4mM + 4i\gamma(q-k)M + 2F_{\alpha\beta}(q_\alpha - k_\alpha)\gamma^\beta\,.$$

Furthermore, we may write

$$\int \bar{u}(q)i(\gamma q)\exp[-i(qx)]\,\mathrm{d}^4q = -\int \bar{u}(q)\cdot m\exp[-i(qx)]\,\mathrm{d}^4q\,,$$

so that

$$\int \mathrm{d}^4q \exp[-iqx]\bar{u}(q)\gamma^\alpha[i\gamma(q-k)-m]M\gamma^\alpha\psi(x) + \text{c.c.}$$
$$= -\int \mathrm{d}^4q\bar{u}(q)\exp[-iqx]\{2[i\gamma k+m]M$$
$$+\,4F_{\alpha\beta}(q_\beta - k_\beta)\gamma^\alpha\}\psi(x) + \text{c.c.}\,.$$

Let

$$II = IIa + IIb\,,$$

where

$$IIa \equiv -\int \mathrm{d}^4q\bar{u}(q)\exp[-iqx]\cdot 2[i\gamma k+m]M\psi(x) + \text{c.c.}\,,$$

$$IIb \equiv -\int \mathrm{d}^4q\bar{u}(q)\exp[-iqx]\cdot 4F_{\alpha\beta}(q_\beta - k_\beta)\gamma^\alpha\psi(x) + \text{c.c.}$$

The first term, IIa, gives another contribution to the magnetic moment; the second term contributes to the *Zitterbewegung*. Furthermore,

$$\bar{D}(k)\frac{\partial \Delta^1(q-k)}{\partial m^2} + D^1(k)\frac{\partial\bar{\Delta}(q-k)}{\partial m^2}$$
$$= \frac{\partial}{\partial m^2}[\bar{D}(k)\Delta^1(q-k) + D^1(k)\bar{\Delta}(q-k)]\Big|_{q=\text{const}}$$
$$= -2\pi\int_0^1 v\cdot\delta''[k^2 + v(q^2+m^2-2kq)]\,\mathrm{d}v\,.$$

Now, we may set $q^2+m^2=0$, and perform the translation $k'=k-vq$. We then again write k' for k, and omit the terms linear in k in the integral because of symmetry reasons. Then,

$$IIa = +\frac{e^3}{2\cdot(2\pi)^3}\int d^4k\int d^4q\, \exp[-iqx]\bar{u}(q)\int_0^1 v\cdot\delta''(k^2+v^2m^2)$$

$$\cdot(i\gamma q\, v+m)M\,\psi(x)\,dv + \text{c.c.}\,.$$

Replacing $i\gamma q$ by $-m$, we obtain

$$IIa = +\frac{e}{4m}\frac{e^2}{4\pi^2}\cdot 2\bar{\psi}M\psi\int_0^1(1-v)\frac{dv}{v}\,. \qquad [29.34]$$

Since $I^{1''}$, IIb, and III are *Zitterbewegungen*, we can already write down the principal result:

$$I^{1'}+IIa = \frac{e}{4m}\cdot\frac{e^2}{4\pi^2}\cdot 2\bar{\psi}M\psi\int_0^1[(1-v)-(1-v^2)]\frac{dv}{v}$$

$$= -\frac{e}{2m}\cdot\frac{e^2}{4\pi^2}\cdot\bar{\psi}M\psi\int_0^1(v-v^2)\frac{dv}{v}\,.$$

The singularity disappears correctly. We have

$$I^{1'}+IIa = -\frac{e}{2m}\cdot\frac{\alpha}{\pi}\cdot\bar{\psi}M\psi\,, \qquad [29.35]$$

where $\alpha=e^2/4\pi\simeq 1/137$. This is the correction to the magnetic moment given in Eq. [29.3].

In addition to the *Zitterbewegung* terms not yet discussed, there still remains term III, which, as we shall see, is also of this form. We gather the terms which we have yet to discuss:

$$I1'' + \text{c.c.} = -\frac{e}{4m}\frac{e^2}{4\pi^2}F_{\alpha\beta}\int \mathrm{d}^4q \exp[-iqx]\bar{u}(q)\gamma^\alpha q_\beta$$

$$\cdot\int_0^1 \frac{(1-v)^2}{v}\psi(x)\,\mathrm{d}v + \text{c.c.}, \qquad [29.36]$$

$$IIb = +\frac{e^3}{(2\pi)^3}F_{\alpha\beta}\int \mathrm{d}^4k\int \mathrm{d}^4q\,\bar{u}(q)\exp[-iqx]\int_0^1 \delta''[k^2 + v(q^2+m^2$$

$$-2kq)]\cdot(q_\beta - k_\beta)\gamma^\alpha\psi(x)v\,\mathrm{d}v + \text{c.c.}. \qquad [29.37]$$

Here, we perform the usual translation $k = k' + vq$, set $q^2 + m^2 = 0$, and drop terms linear in k'. Then,

$$IIb = \frac{e^3}{(2\pi)^2}\frac{1}{2m^2}F_{\alpha\beta}\int \mathrm{d}^4q \exp[-iqx]\bar{u}(q)\gamma^\alpha q_\beta$$

$$\cdot\int_0^1 \frac{1-v}{v}\psi(x)\,\mathrm{d}v + \text{c.c.}. \qquad [29.38]$$

There still remains Eq. [29.26],

$$III = i\frac{e^3}{8}F_{\mu\nu}\int \mathrm{d}^4q \exp[-iqx]\bar{u}(q)\gamma^\alpha\int \mathrm{d}^4\xi \exp[-iq\xi]$$

$$\cdot[\bar{D}(\lambda)\Delta^1(\lambda) + D^1(\lambda)\bar{\Delta}(\lambda)]\gamma^\mu\gamma^\alpha\xi^\nu\psi(x) + \text{c.c.}.$$

Because of $\gamma^\alpha\gamma^\mu\gamma^\alpha = -2\gamma^\mu$ and our lemma of Eq. [29.19],

$$\int \exp[-iq\xi][\bar{D}(\lambda)\Delta^1(\lambda) + D^1(\lambda)\bar{\Delta}(\lambda)]\,\mathrm{d}^4\xi$$

$$= -\left(\frac{1}{2\pi}\right)^3\int \mathrm{d}^4k\int_0^1 \delta'\big(k^2 + v(q^2+m^2-2kq)\big)\,\mathrm{d}v,$$

which, upon differentiation with respect to q_ν yields

$$\int \exp[-iq\xi]\xi^\nu[\bar{D}(\lambda)\Delta^1(\lambda) + D^1(\lambda)\bar{\Delta}(\lambda)]\,d^4\xi$$

$$= \left(\frac{1}{2\pi}\right)^3 \cdot i \frac{\partial}{\partial q_\nu} \int d^4k \int_0^1 \delta'\big(k^2 + v(q^2 + m^2 - 2kq)\big)\,dv$$

$$= -\frac{i}{(2\pi)^3} \int d^4k \cdot 2\,(q_\nu - k_\nu) \int_0^1 \delta''\big(k^2 + v(q^2 + m^2 - 2kq)\big) v\,dv\,,$$

and we obtain

$$III = -\frac{1}{2}\frac{e^3}{(2\pi)^3} F_{\mu\nu} \int d^4q \exp[-iqx]\,\bar{u}(q) \int d^4k (q_\nu - k_\nu)$$

$$\cdot \int_0^1 \delta''\big(k^2 + v(q^2 + m^2 - 2kq)\big)\psi(x) v\,dv + \text{c.c.}\,, \qquad [29.39]$$

$$III = -\tfrac{1}{2} IIb\,,$$

so that

$$III + IIb = +\frac{1}{2} IIb = +\frac{e}{4m^2}\frac{e^2}{(2\pi)^2} F_{\alpha\beta}$$

$$\cdot \int d^4q \exp[-iqx]\,\bar{u}(q)\gamma^\alpha q_\beta \int_0^1 \psi(x)\frac{1-v}{v}\,dv + \text{c.c.}\,.$$

Now, with

$$I^{1''} = -\frac{e}{4m^2}\frac{e^2}{(2\pi)^2} F_{\alpha\beta} \int d^4q \exp[-iqx]\bar{u}(q)\,\gamma^\alpha q_\beta$$

$$\cdot \int_0^1 \psi(x)\frac{(1-v)^2}{v}\,dv + \text{c.c.}\,,$$

we define

$$Z \equiv I^{1''} + IIb + III = \frac{e^2}{4\pi^2}\cdot\frac{e}{4m^2} F_{\alpha\beta} \int d^4q\,\bar{u}(q) \exp[-iqx]\,\gamma^\alpha q^\beta$$

$$\cdot \int_0^1 \psi(x)\,[1 - (1 - v)]\,\frac{1-v}{v}\,dv + \text{c.c.}\,,$$

$$Z = \frac{e^2}{4\pi^2}\frac{e}{4m^2}\cdot\frac{1}{2}F_{\alpha\beta}\int d^4q\,\bar{u}(q)\exp[-iqx]\,\gamma^\alpha q_\beta\psi(x) + \text{c.c.}$$

$$= \frac{e^2}{4\pi^2}\cdot\frac{e}{8m^2}F_{\alpha\beta}\left\{i\frac{\partial\bar{\psi}}{\partial x^\beta}\gamma^\alpha\psi(x) + \text{c.c.}\right\}.$$

Since

$$ie\bar{\psi}\gamma^\alpha\psi \equiv j^\alpha,$$

then

$$\text{c.c.}\,(i\bar{\psi}\gamma^\alpha\psi) = +\,i\bar{\psi}\gamma^\alpha\psi\,.$$

Therefore,

$$Z = \frac{e^2}{4\pi^2}\frac{e}{8m^2}\cdot iF_{\alpha\beta}\left\{\frac{\partial\bar{\psi}}{\partial x^\beta}\gamma^\alpha\psi(x) + \bar{\psi}\gamma^\alpha\frac{\partial\psi}{\partial x^\beta}\right\}$$

$$= \frac{e^2}{4\pi^2}\cdot\frac{e}{8m^2}\cdot F_{\alpha\beta}\frac{\partial}{\partial x^\beta}(i\bar{\psi}\gamma^\alpha\psi)$$

$$= \frac{\alpha}{\pi}\frac{1}{8m^2}F_{\alpha\beta}\frac{\partial j^\alpha}{\partial x^\beta},$$

$$Z = \frac{\alpha}{\pi}\frac{1}{8m^2}\frac{\partial}{\partial x^\beta}(F_{\alpha\beta}j^\alpha),$$

since $F_{\alpha\beta}$ is constant. Alternatively,

$$Z = \frac{\alpha}{\pi}\frac{1}{16m^2}F_{\alpha\beta}\left(\frac{\partial j^\alpha}{\partial x^\beta} - \frac{\partial j^\beta}{\partial x^\alpha}\right). \qquad [29.40]$$

Indeed, such terms when integrated only over space do not give zero, but instead a result of the form $\boldsymbol{E}\cdot(\mathrm{d}\boldsymbol{J}/\mathrm{d}t)$.

However, the time integral of this expression vanishes. Thus, these terms have only matrix elements which correspond to a *Zitterbewegung*, and which give, as stated, no contribution to the self-energy.

Final Remark: The advantage of the Villars-Géhéniau method for the calculation of the anomalous magnetic moment lies principally in the fact that no charge renormalization terms appear. This must physically be so since the sources of the field are at infinity. On the contrary, Schwinger's method yields many such charge renormalization terms which, in the end, cancel.

Chapter 7. Feynman's Approach to Quantum Electrodynamics [1]

30. THE PATH INTEGRAL METHOD

Feynman begins with a formulation of quantum mechanics that avoids the Hamiltonian. The origin of this procedure is a remark by Dirac,[2] which Feynman has developed.[3] See also Choquard.[4]

We start from a special solution of the Schrödinger equation for n degrees of freedom,

$$\frac{\hbar}{i}\frac{\partial \psi}{\partial t} + \underline{H}\psi = 0 , \qquad [30.1]$$

of the form

$$\psi(q,t) = K(q,t;\, q',t') , \qquad [30.2]$$

which, for $t=t'$, reduces to the n-dimensional δ-function:

$$K(q,t;\, q',t) = \delta^{(n)}(q-q') . \qquad [30.3]$$

This K, then, gives us the solution for any arbitrary initial state, $\psi(q',t')$:

$$\psi(q,t) = \int K(q,t;\, q',t')\,\psi(q',t')\,\mathrm{d}^n q' . \qquad [30.4]$$

[1] R. P. Feynman, *Phys. Rev.* **76**, 769 (1949); **80**, 440 (1950).

[2] P. A. M. Dirac, *The Principles of Quantum Mechanics* (Oxford University Press, Oxford, 1947), 3rd ed., Sect. 32, p. 125, "The action principle."

[3] R. P. Feynman, *Rev. Mod. Phys.* **20**, 367 (1948).

[4] Ph. Choquard, *Helv. Phys. Acta* **28**, 89 (1955).

The probability amplitude that the system goes from the state ψ_n to the state ψ_m during the time τ is given by the following matrix element:

$$K_{mn} \equiv \int \mathrm{d}^n q \int \mathrm{d}^n q' \psi_m^*(q, t+\tau) K(q, t+\tau; q', t) \psi_n(q', t) .$$

For simplicity, we assume in what follows that $\underline{H}$ does not depend explicitly upon time; then, K contains only the time difference $\tau = t - t'$:

$$K(q, t; q', t') = K(q, \tau; q', 0) . \qquad [30.5]$$

This restriction, however, is not essential.

Properties of K:

1. $$\int K(q, q'; \tau) K^*(q, q''; \tau) \mathrm{d}^n q = \delta^{(n)}(q' - q'') . \qquad [30.6]$$

This is true because, from the continuity equation, it follows that the integral is time-independent. For $\tau = 0$, however, it is equal to the right-hand side of the above equation.

2. $$K^*(q', q; \tau) = K(q, q'; -\tau) . \qquad [30.7]$$

This follows immediately from the Hermiticity of $\underline{H}$.

3. K has the group property

$$\int K(q, q'; \tau_1) K(q', q''; \tau_2) \mathrm{d}^n q' = K(q, q''; \tau_1 + \tau_2) , \qquad [30.8]$$

since (*a*) the Schrödinger equation (in q) is satisfied by both sides, and (*b*) the initial condition [30.3] is satisfied, which follows from properties 1 and 2.

Feynman has attempted to give a new foundation of quantum mechanics in that he omits the Schrödinger equation (Eq. [30.1]),

$$\frac{\hbar}{i} \frac{\partial K}{\partial t} + \underline{H} K = 0 \qquad \text{(note: } \underline{H} \text{ operates only upon } q) ,$$

and in its place introduces K axiomatically. In addition, Eq. [30.4] is to hold.

As is well known, the Schrödinger equation permits a transition according to the correspondence principle from a classical problem $(H(p, q))$ to the related quantum mechanical one $(\underline{H}(\underline{p}, \underline{q}))$, aside from the ambiguities in the ordering of the factors. Something analogous is possible here. One can find a solution $K(q, q'; \tau)$ *for small* τ. One defines

$$K_0(q, q'; \tau) = (2\pi i\hbar)^{-n/2} \sqrt{D} \exp\left[\frac{i}{\hbar} S(q, q'; \tau)\right]. \quad [30.9]$$

Here, $S(q, q'; \tau)$ is the classical action integral

$$S(q, q'; \tau) = \int_t^{t+\tau} L\,\mathrm{d}t', \quad [30.10]$$

where the integral is to extend along the classical path from q' to q. If $H(p, q)$ depends explicitly upon time, one must write

$$S(q, t+\tau; q', t) = \int_t^{t+\tau} L\,\mathrm{d}t'. \quad [30.11]$$

Furthermore,

$$D = (-1)^n \left\| \frac{\partial^2 S}{\partial q^i \partial q'^k} \right\|, \quad [30.12]$$

where $\| \ \|$ represents the determinant. According to classical mechanics,

$$p_k = \frac{\partial S}{\partial q^k}, \qquad p'_k = -\frac{\partial S}{\partial q'^k}, \quad [30.13]$$

and the Hamilton-Jacobi equations hold:

$$\left.\begin{aligned} \frac{\partial S}{\partial \tau} + H\left(\frac{\partial S}{\partial q}, q\right) = 0 \\ \frac{\partial S}{\partial \tau} + H\left(-\frac{\partial S}{\partial q'}, q'\right) = 0 \end{aligned}\right\}. \quad [30.14]$$

We will now derive a differential equation for K_c. This will, in general, not be the Schrödinger equation—which really ought not to be expected, for otherwise wave mechanics would be almost superfluous since K_c is, indeed, constructed from purely classical quantities. However, for small τ this differential equation goes over into the Schrödinger equation. That K_c for $\tau = 0$ goes over into the δ-function is more critical, and will, therefore, be explained with examples.

We will restrict ourselves to a "normal" Lagrangian function containing a magnetic field and written in Cartesian coordinates:

$$\left.\begin{aligned} L &= \sum_k \frac{m_k}{2}\left((\dot{q}^k)^2 - \dot{q}^k A_k(q)\right) - V(q) \\ H &= \sum_k \frac{1}{2m_k}\left(p_k + A_k(q)\right)^2 + V(q) \end{aligned}\right| . \qquad [30.15]$$

(The case of curvilinear coordinates can be carried out in exactly the same manner.) Thus,

$$\frac{\partial S}{\partial \tau} + \sum_k \frac{1}{2m_k}\left(\frac{\partial S}{\partial q^k} + A_k(q)\right)^2 + V(q) = 0 . \qquad [30.16]$$

We now need $\partial D/\partial \tau$ and, therefore,

$$\frac{\partial}{\partial \tau}\left(\frac{\partial^2 S}{\partial q^i \partial q'^k}\right).$$

Equation [30.16] implies

$$\frac{\partial}{\partial \tau}\frac{\partial S}{\partial q'^i} + \sum_k \frac{1}{m_k}\left(\frac{\partial S}{\partial q^k} + A_k\right)\frac{\partial^2 S}{\partial q^k \partial q'^i} = 0 .$$

Taking $\partial/\partial q^j$ of this expression gives

$$\frac{\partial}{\partial \tau}\frac{\partial^2 S}{\partial q'^i \partial q^j} + \sum_k \frac{1}{m_k}\frac{\partial}{\partial q^j}\left(\frac{\partial S}{\partial q^k} + A_k\right)\frac{\partial^2 S}{\partial q'^i \partial q^k}$$
$$+ \sum_k \frac{1}{m_k}\left(\frac{\partial S}{\partial q^k} + A_k\right)\frac{\partial^3 S}{\partial q^j \partial q'^i \partial q^k} = 0 . \qquad [30.17]$$

Let

$$\varphi_{ji} = \frac{\partial^2 S}{\partial q^j \partial q'^i}, \qquad (\varphi_{ji} \neq \varphi_{ij}),$$

and define φ^{ji} by

$$\sum_\alpha \varphi^{j\alpha} \varphi_{\alpha i} = \delta^j_i = \sum_\alpha \varphi^{\alpha j} \varphi_{i\alpha}.$$

Then,

$$\frac{\partial D}{\partial \tau} = D \cdot \varphi^{ji} \frac{\partial \varphi_{ij}}{\partial \tau}$$

and

$$\frac{\partial D}{\partial q^k} = D \cdot \varphi^{ji} \frac{\partial \varphi_{ij}}{\partial q^k},$$

where the summation convention is employed. With this, upon multiplying Eq. [30.17] by φ^{ji} and contracting, there follows

$$\frac{1}{D}\frac{\partial D}{\partial \tau} + \sum_k \frac{1}{m_k} \frac{\partial}{\partial q^k}\left(\frac{\partial S}{\partial q^k} + A_k\right) + \sum_k \frac{1}{m_k}\left(\frac{\partial S}{\partial q^k} + A_k\right)\frac{1}{D}\frac{\partial D}{\partial q^k} = 0. \qquad [30.18]$$

We now form

$$\frac{\hbar}{i}\frac{\partial K_C}{\partial \tau} + \left(\sum_k \frac{1}{2m_k}\left(\frac{\hbar}{i}\frac{\partial}{\partial q^k} + A_k\right)^2 + V(q)\right) K_C. \qquad [30.19]$$

We have

$$\left(\frac{\hbar}{i}\frac{\partial}{\partial q^k} + A_k\right) K_C = \left(\frac{\partial S}{\partial q^k} + A_k + \frac{\hbar}{i}\frac{1}{2D}\frac{\partial D}{\partial q^k}\right) K_C,$$

$$\left(\frac{\hbar}{i}\frac{\partial}{\partial q^k} + A_k\right)^2 K_C = \left(\frac{\partial S}{\partial q^k} + A_k + \frac{\hbar}{i}\frac{1}{2D}\frac{\partial D}{\partial q^k}\right)^2 K_C + \frac{\hbar}{i}\left[\frac{\partial}{\partial q^k}\left(\frac{\partial S}{\partial q^k} + A_k + \frac{\hbar}{i}\frac{1}{2D}\frac{\partial D}{\partial q^k}\right)\right] \cdot K_C,$$

$$\frac{\hbar}{i}\frac{\partial K_C}{\partial \tau} = \left(\frac{\partial S}{\partial \tau} + \frac{1}{2}\frac{\hbar}{i}\frac{1}{D}\frac{\partial D}{\partial \tau}\right) K_C.$$

We order according to powers of $\hbar$ and add. Then, the zero-order terms in $\hbar$ vanish because of the Hamilton-

Jacobi equation, Eq. [30.16], and the first-order terms in $\hbar$ yield

$$\frac{\hbar}{2i}\left[\frac{1}{D}\frac{\partial D}{\partial \tau}+\sum_k \frac{1}{m_k}\frac{\partial}{\partial q^k}\left(\frac{\partial S}{\partial q^k}+A_k\right)\right.$$
$$\left.+\sum_k \frac{1}{m_k}\left(\frac{\partial S}{\partial q^k}+A_k\right)\frac{1}{D}\frac{\partial D}{\partial q^k}\right]=0\,. \qquad [30.20]$$

There remain only the terms in $\hbar^2$ (the "false terms"):

$$-\hbar^2\left(\sum_k \frac{1}{2m_k}\left[\left(\frac{1}{2D}\frac{\partial D}{\partial q^k}\right)^2+\frac{\partial}{\partial q^k}\left(\frac{1}{2D}\frac{\partial D}{\partial q^k}\right)\right]\right)\cdot K_C\,.$$

If these are combined, then there follows

$$\frac{\hbar}{i}\frac{\partial K_C}{\partial \tau}+\underline{H}K_C=-\hbar^2\left(\sum_k \frac{1}{2m_k}\frac{1}{\sqrt{D}}\frac{\partial^2\sqrt{D}}{\partial (q^k)^2}\right)\cdot K_C\,.$$

Note: If D is independent of q, then $K_C=K$ is already the correct solution.

We now wish to show that in the general case, although K_C is not the desired K, *for small* τ it does have the right properties. More specifically, we wish to show that

1. $$K_C(q,q';0)=\delta^{(n)}(q-q')\,,$$

2. $$\lim_{\tau\to 0}\frac{K-K_C}{\tau}=0\,.$$

If these are satisfied, then, with the aid of the group property [30.8] we can obtain the proper K from K_C by means of a limiting process:

$$K(q,q';\tau)=\lim_{\substack{\varepsilon\to 0\\ N\to\infty\\ \tau\ \text{fixed}}}\int\prod_{\alpha=0}^{N-1}K_C(q^{(\alpha+1)},q^{(\alpha)};\varepsilon)\,dq^1\ldots dq^\alpha\ldots dq^{N-1}, \qquad [30.21]$$

in which we have divided τ into N intervals of size ε,

$$\tau=N\varepsilon\,,\quad \text{and}\quad q'=q^0,\quad q=q^N$$

and K_C is given by [30.9]. Equation [30.21] can be interpreted as an integral over all classical paths.[5] Because

[5] R. P. FEYNMAN, *Rev. Mod. Phys.* **20**, 367 (1948).

$\lim\limits_{\tau \to 0} (K - K_c)/\tau = 0$, the error in K cannot give a finite result in Eq. [30.21], so that this is in fact the correct K.

We must still prove assertions 1 and 2. To this end, we consider a few examples.

a. The free fall:

$$L = \frac{m}{2} \dot{q}^2 + mgq,$$

$$K_c(q, q'; \tau) = K(q, q'; \tau) = \sqrt{\frac{m}{2\pi i\hbar\tau}}$$

$$\cdot \exp\left[\frac{i}{\hbar} m \left(\frac{(q' - q)^2}{2\tau} + \frac{\tau}{2} g(q + q') - \frac{1}{24} g^2 \tau^3\right)\right]. \qquad [30.22]$$

b. The linear harmonic oscillator:

$$L = \frac{m}{2} \dot{q}^2 - \frac{m\omega^2}{2} q^2,$$

$$K_c(q, q'; \tau) = K(q, \cdot q'; \tau)$$

$$= \sqrt{\frac{m\omega}{2\pi i\hbar \sin \omega\tau}} \exp\left[\frac{i}{\hbar} m\omega \frac{(q^2 + q'^2) \cos \omega\tau - 2qq'}{2 \sin \omega\tau}\right]. \qquad [30.23]$$

c. A free particle in a homogeneous magnetic field:

$$L = \frac{m}{2} (\dot{q}_1^2 + \dot{q}_2^2) - m\sigma(\dot{q}_1 q_2 - \dot{q}_2 q_1), \qquad \sigma = \frac{eH}{2mc},$$

$$K_c(\mathbf{q}, \mathbf{q}'; \tau) = K(\mathbf{q}, \mathbf{q}'; \tau) = \frac{m\sigma}{2\pi i\hbar \sin \sigma\tau}$$

$$\cdot \exp\left[\frac{i}{\hbar} m\sigma \left(\frac{(\mathbf{q} - \mathbf{q}')^2 \cos \sigma\tau}{2 \sin \sigma\tau} + q_2' q_1 - q_1' q_2\right)\right]. \qquad [30.24]$$

For the proof of the above assertions, we write, corresponding to our expectation,

$$\int K(q, q'; \tau) \varphi(q') \, dq' = \varphi(q) + \frac{i}{\hbar} \tau \underline{H} \varphi(q) + \tau \chi(q, \tau), \qquad [30.25]$$

and wish to show that

$$\lim_{\tau \to 0} \chi(q, \tau) = 0. \qquad [30.26]$$

This includes both assertions 1 and 2. For the case of example a, the calculation can be carried out exactly. In doing this, all the essentials can already be seen. We have

$$\int K(q, q'; \tau)\varphi(q')\,\mathrm{d}q' \tag{30.27}$$

$$= \sqrt{\frac{m}{2\pi i\hbar\tau}}\int \exp\left[\frac{i}{\hbar}m\left(\frac{(q'-q)^2}{2\tau} + \frac{\tau}{2}g(q+q') - \frac{1}{24}g^2\tau^3\right)\right]\varphi(q')\,\mathrm{d}q',$$

$$= \exp[-i\tau gq]\sqrt{\frac{m}{2\pi i\hbar\tau}}\int \exp\left[\frac{i}{\hbar}\frac{m}{2\tau}\left(\xi^2 - \tau^2 g\xi - \frac{1}{12}g^2\tau^4\right)\right] \cdot\varphi(q+\xi)\,\mathrm{d}\xi.$$

With $\xi - \frac{1}{2}g\tau^2 = u$, Eq. [30.27] is equal to

$$= \exp[-i\tau gq]\sqrt{\frac{m}{2\pi i\hbar\tau}}\int \exp\left[\frac{i}{\hbar}\frac{m}{2\tau}\left(u^2 + 2\tau^2 gq - \frac{1}{12}g^2\tau^4 + \frac{1}{4}g^2\tau^4\right)\right]\cdot\varphi(q + u + \tfrac{1}{2}g\tau^2)\,\mathrm{d}u\,.$$

If $u = v\sqrt{2\tau}$, this becomes

$$= \exp[-i\tau gq]\sqrt{\frac{m}{\pi i\hbar}}\int \exp\left[\frac{i}{\hbar}m[v^2 + O(\tau^2)]\right] \cdot\varphi(q + \sqrt{2\tau}v + \tfrac{1}{2}g\tau^2)\,\mathrm{d}v.$$

With

$$\int_{-\infty}^{+\infty} \exp[iau^2]\,\mathrm{d}u = \sqrt{\frac{i\pi}{a}} \qquad \left(\sqrt{i} \equiv \exp\left[+i\frac{\pi}{4}\right]\right),$$

$$\int_{-\infty}^{+\infty} \exp[iau^2]\,u\cdot\mathrm{d}u = 0\,,$$

$$\int_{-\infty}^{+\infty} \exp[iau^2]u^2\,\mathrm{d}u = \frac{1}{2}\left(\frac{-i\pi}{a^3}\right)^{\frac{1}{2}},$$

and

$$\varphi(q+\sqrt{2\tau}v+\tfrac{1}{2}g\tau^2)=\varphi(q)+\sqrt{2\tau}v\varphi'(q)+\tau v^2\varphi''(q)+O(\tau^2),$$

there results

$$\int K(q,q';\tau)\varphi(q')\,\mathrm{d}q'=\exp\left[-\frac{i\tau gq}{\hbar}\right]\cdot[\varphi(q)+\tau\cdot i\hbar\varphi''(q)]+O(\tau^2) \qquad [30.28]$$

$$=\varphi(q)-\frac{i\tau}{\hbar}\left[-\frac{\hbar^2}{2m}\frac{\partial^2}{\partial q^2}+gq\right]\varphi(q)+O(\tau^2),$$

which is what we wished to show. For example b, the procedure goes exactly the same way. The general proof is patterned after this calculation. The idea is as follows.

In the somewhat unusual variables q and q', the smaller $|q-q'|$ is, the better one can replace $S(q, q'; \tau)$ at fixed τ by the force-free S.

Force-free case:

$$S_0(q,q';\tau)=m\frac{(q'-q)^2}{2\tau}.$$

General case:

$$S(q,q';\tau)=m\frac{(q'-q)^2}{2\tau}-\tau V(q)+S_1(q,q';\tau),$$

where $S_1(q, q'; \tau)\to 0$ for $q'\to q$ at fixed τ. $\partial S_1/\partial q$ gives the change in the final momentum which is a consequence of the force acting along the path. Let $F(q)$ be the force. Then, physically, our assertion is that if

$$|F(q+\lambda q')|\tau \ll m\frac{|q'-q|}{\tau} \qquad [30.29]$$

(that is, if the action of the force is much less than the momentum in the force-free case), then

$$|S_1|< C\frac{\tau^2|F|}{m|q'-q|}\cdot|S_0|. \qquad [30.30]$$

If one limits oneself to not too rapidly increasing forces (i.e., if $|F|/|q'-q|$ is bounded), then $|S_1|$ for large $|q'-q|$ is uniformly bounded. Then, the integral can be evaluated in the following manner. There exists a ξ_0 such that for $|q'-q| > \xi_0$ our above condition is satisfied, so that an error $\leqslant O(\tau^2)$ is made if S_1 is neglected. For $\xi < \xi_0$ this is not permitted, but ξ_0 can be so chosen that $\lim_{\tau\to 0}|(1/\tau)\xi_0| = 0$, so that the contribution of this integral also vanishes because the integrand is, indeed, bounded. This ξ_0 exists because it must be true that

$$\frac{m}{1} F(q)\cdot\tau^2 \ll \xi_0 \ll \sqrt{\frac{2\tau\hbar}{m}}\,.$$

The best ξ_0 is

$$\xi_0 = \tau^{5/4}\frac{F}{\sqrt{2m\hbar}}\,,$$

and this, in fact, satisfies all conditions. With this it is, therefore, shown that

$$\int K(q, q'; \tau)\varphi(q')\,\mathrm{d}q' = \varphi(q) - \frac{\hbar}{i}\tau\underline{H}\varphi(q) + \chi(q, \tau)\,, \qquad [30.31]$$

where

$$\lim_{\tau\to 0}\frac{\chi(q, \tau)}{\tau} = 0\,.$$

Note: 1. The same calculation can also be carried through for the case of velocity-dependent forces (as in the presence of a magnetic field).

2. The only restriction is that

$$|F(q)| < M\cdot|q|\,.$$

That this condition for being able to use the force-free K_0 for small τ and fixed $|q'-q|$ is also necessary, can be seen in the example—also interesting in its own right—of a particle in a one-dimensional box of length L. This corresponds to a potential

$$V(q) = C \cdot \left(\frac{|q|}{L}\right)^{\infty} .$$

Then,

$$\frac{\hbar}{i}\frac{\partial \psi}{\partial t} + \frac{p^2}{2m}\psi = 0 , \qquad \text{where} \qquad \psi(0, t) = \psi(L, t) = 0 .$$

With

$$\psi_n(q, t) = u_n(q) \exp[-i\nu_n t] ,$$

we have

$$u_n(q) = \sqrt{\frac{2}{L}} \sin\left(\frac{n\pi q}{L}\right),$$

$$\nu_n = \frac{1}{\hbar}\frac{p_n^2}{2m} = \frac{\hbar\pi^2}{2mL^2} n^2 .$$

The kernel K can be obtained in two ways:

1. From the familiar representation of the Green's function in terms of eigenfunctions,

$$K(q, q'; \tau) = \sum_n u_n^*(q) u_n(q') \exp[-i\nu_n \tau] .$$

Then,

$$K(q, q'; \tau) = \frac{1}{L}\sum_{n=1}^{\infty}\left[\cos\left(\frac{n\pi}{L}(q - q')\right) - \cos\left(\frac{n\pi}{L}(q + q')\right)\right] \cdot \exp\left[-i\frac{\hbar\pi^2\tau}{2mL^2} n^2\right] . \qquad [30.32]$$

2. From the kernel for the free particle,

$$K_0(q, q'; \tau) = \sqrt{\frac{m}{2\pi i \hbar \tau}} \exp\left[\frac{im}{2\hbar\tau}(q - q')^2\right] \equiv K_0(q - q') ,$$

using the method of images,

$$K'(q, q'; \tau) = \sum_{n=-\infty}^{+\infty} [K_0(q - q' - 2Ln) - K_0(q + q' - 2Ln)] .$$

Both kernels K and K' are, of course, identical, which

can most easily be seen as follows. In the notation of Whittaker and Watson,[6]

$$\vartheta_3(z|\tau) = \sum_{n=-\infty}^{+\infty} \exp[2niz] \exp[i\pi\tau n^2]$$

$$= 1 + 2 \sum_{n=1}^{\infty} \cos 2nz \cdot \exp[i\pi\tau n^2],$$

and it is true that

$$\vartheta_3(z|\tau) = \vartheta_3(-z|\tau)$$

and

$$\vartheta_3(z|\tau) = (-i\tau)^{-\frac{1}{2}} \exp\left[\frac{z^2}{i\pi\tau}\right] \cdot \vartheta_3\left(\frac{z}{\tau}\middle| -\frac{1}{\tau}\right). \qquad [30.33]$$

There then results

$$K(q, q'; \tau) = \frac{1}{2L}\left[\vartheta_3\left(\frac{\pi(q-q')}{2L}\middle| -\frac{\hbar\pi\tau}{2mL^2}\right) - \vartheta_3\left(\frac{\pi(q+q')}{2L}\middle| -\frac{\hbar\pi\tau}{2mL^2}\right)\right]$$

and

$$K'(q, q'; \tau) = \sqrt{\frac{m}{2\pi i\hbar\tau}}\left\{\exp\left[\frac{im}{2\hbar\tau}(q-q')^2\right] \cdot \vartheta_3\left(+\frac{mL}{\hbar\tau}(q-q')\middle|\frac{2m}{\pi\hbar\tau}L^2\right) - \exp\left[\frac{im}{2\hbar\tau}(q+q')^2\right] \cdot \vartheta_3\left(+\frac{mL}{\hbar\tau}(q+q')\middle|\frac{2m}{\pi\hbar\tau}L^2\right)\right\},$$

and, because of relation [30.33], these two expressions reduce to the same thing.

In this example, one recognizes that for $|F(q)| > M \cdot |q|$, the force-free kernel K_0 need not yield a good approximation for small τ and fixed $|q'-q|$. Here, nervertheless, the classical K_c is a rigorous solution of the Schrödinger equation.

[6] E. T. Whittaker and G. N. Watson, *A Course of Modern Analysis* (Cambridge University Press, Cambridge, 1927), 4th ed., p. 462.

Note: The ϑ_3 series converges uniformly only for $\operatorname{Im}\tau > 0$; actually, the kernels we have written down do not exist for real τ. Nevertheless, it can be shown that

$$\int K(q, q'; \tau)\varphi(q')\,\mathrm{d}q'$$

exists for sufficiently "normal" functions $\varphi(q')$, and that it is given by

$$\lim_{\varepsilon \to 0} \int K(q, q'; \tau + i\varepsilon)\varphi(q')\,\mathrm{d}q' .$$

All of the above formulas are to be considered in this sense, then—as the limiting value for the vanishing imaginary part after having performed the integration.

With the aid of this formalism, then, Feynman also introduces quantum electrodynamics. First of all, one can give a rigorous solution for the forced oscillations of a harmonic oscillator.[7]

$$\left.\begin{aligned} L &= \tfrac{1}{2}(\dot{q}^2 - \omega^2 q^2) + \gamma(t)\cdot q \\ H &= \tfrac{1}{2}(p^2 + \omega^2 q^2) - \gamma(t)\cdot q \end{aligned}\right\}, \qquad [30.34]$$

so that

$$\ddot{q} + \omega^2 q = \gamma(t) .$$

Note: Here, $K(q, t; q', t')$ is not dependent only on $\tau = t - t'$; nevertheless, this changes nothing essential in the previous considerations.

We have then [7]

$$S = \frac{\omega}{2\sin\omega\tau}\Bigg[(q^2 + q'^2)\cos\omega\tau - 2qq'$$

$$+ \frac{2q}{\omega}\int_{t'}^{t}\gamma(u)\sin\omega(u - t')\,\mathrm{d}u + \frac{2q'}{\omega}\int_{t'}^{t}\gamma(v)\sin\omega(t - v)\,\mathrm{d}v$$

$$+\left(-\frac{2}{\omega^2}\right)\int_{t'}^{t}\mathrm{d}v\int_{t'}^{v}\mathrm{d}u\,\gamma(v)\gamma(u)\sin\omega(t - v)\sin\omega(u - t')\Bigg]. \qquad [30.35]$$

[7] R. P. Feynman, *Phys. Rev.* **80**, 440 (1950); Section 3.

In particular, the diagonal element of the ground state is

$$K_{00}(t,t') = \int dq \int dq' \psi_0^*(q,t) K(q,t;q',t') \psi_0(q',t') ,$$

$$\psi_0(q,t) = \sqrt[4]{\frac{\omega}{\pi}} \exp\left[-\frac{1}{2}\omega q^2 - \frac{i}{2}\omega t\right] ,$$

$$K_{00}(t,t') = \exp\left[-\frac{1}{2\omega} \cdot \int_{t'}^{t} dv \int_{t'}^{v} du\, \gamma(u)\gamma(v) \exp[-i\omega(v-u)]\right] , \quad [30.36]$$

as can easily be verified.[8]

With this, the field oscillators can be eliminated. Let us consider a system—an electron, for example—coupled to an oscillator. Then,

$$H = H_0(x) + \tfrac{1}{2}(p^2 + \omega^2 q^2) - \gamma[x(t)] \cdot q . \quad [30.37]$$

If we then apply our product expression [30.21] for K and, in particular, ask for the matrix element for which the oscillator remains in the ground state ($\sim$ photon vacuum), then we can integrate over q to find

$$K(x,x';\tau) = \lim_{\varepsilon \to 0} \int \dots \int \prod_{\alpha=0}^{N} \sqrt{D} \exp\left[\frac{i}{\hbar} S_0(x_{\alpha+1}, x_\alpha; \varepsilon)\right] \cdot K_{00}(\gamma(x_1) \dots \gamma(x_n))\, d^N x ,$$

where $\tau = (N+1)\varepsilon$, $x' = x_0$, $x = x_{N+1}$, and K_{00} is to be considered a functional of $\gamma(x)$; or, if the integral is written as a Riemann sum, then it is to be considered as a function of the N quantities $\gamma(x_\alpha)$.

Feynman then applies this to the infinity of field oscillators and obtains exactly Dyson's formulas, as must of course be the case. In particular, the D^C function comes out very nicely.

As one can see here, the formalism is exactly equivalent to the usual quantum electrodynamics. However, in this manner Feynman derived the Dyson formulas before Dyson did.

[8] Ibid.

Supplementary Bibliography

A. Akhiezer and V. B. Bereztetski, *Quantum Electrodynamics* (Wiley, New York, 1963).

J. D. Bjorken and S. D. Drell, *Relativistic Quantum Fields* (McGraw-Hill, New York, 1965).

N. N. Bogoliubov and D. V. Shirkov, *Introduction to the Theory of Quantized Fields* (Interscience, New York, 1959).

G. F. Chew, *S-Matrix Theory of Strong Interactions* (Benjamin, New York, 1962).

R. P. Feynman, *Quantum Electrodynamics* (Benjamin, New York, 1962).

A. O. G. Källén, "Quantenelektrodynamik" in *Encyclopedia of Physics* (S. Flügge, ed.), Vol. V, Part 1 (Springer, Berlin, 1958).

F. Mandl, *Introduction to Quantum Field Theory* (Interscience, New York, 1960).

S. Schweber, *An Introduction to Relativistic Quantum Field Theory* (Harper and Row, New York, 1961).

J. Schwinger, *Quantum Electrodynamics* (Dover, New York, 1958).

Appendix. Comments by the Editor

[A-1] (p. 1). The following *notation* is used:

Quantities a^μ, b_μ with Greek index $\mu = 1, 2, 3, 4$ designate 4-vectors. Scalar products are written with the usual summation convention as

$$(ab) = a^\mu b_\mu = \boldsymbol{a} \cdot \boldsymbol{b} - a^0 b_0$$

where $\boldsymbol{a}$, $\boldsymbol{b}$ are 3-vectors and $a^4 = ia^0$, $b_4 = ib_0$. In this notation there is no distinction between upper and lower indices. Arbitrarily, upper indices are used for position vectors x^μ ($x^0 = t$), currents j^μ, and Dirac matrices γ^μ (the latter are all Hermitian); lower indices for momentum vectors k_μ, quantized vector potentials Φ_μ, and external (c-number) vector potentials $\mathscr{A}_\mu$. Furthermore $d^4x = d^3x\,dx_0$, $d^4k = d^3k\,dk_0$.

An upper index $*$ is used for complex conjugation of c-numbers as well as for Hermitian conjugation of q-numbers. Exceptions are Eq. [3.13] where the star means complex conjugation, and Eq. [15.10] where the meaning of the star is defined in Eq. [15.15]. In the last equations as well as in other places such as in Eqs. [3.17] to [3.19] Hermitian conjugation is designated by an upper index †. No attempt to correct these inconsistencies has been made since the meaning is evident in each instance.

Units are used such that $\hbar = c = 1$. Maxwell's equations are written in Heaviside units such that $e^2/(4\pi) = \alpha \cong 1/137$, e being the charge of the electron and α the fine structure constant. Then

$$\Box\Phi_\mu = -j^\mu\,; \qquad \frac{\partial\Phi_\mu}{\partial x^\mu} = 0\,,$$

where

$$\Box = \frac{\partial}{\partial x^\mu}\frac{\partial}{\partial x^\mu} = \nabla^2 - \frac{\partial^2}{\partial t^2}$$

is the d'Alembertian and the second equation is the Lorentz condition. For details see W. Pauli, *Electrodynamics* (M.I.T. Press, Cambridge, Mass., 1972).

[A-2] (p. 48). An interpretation in terms of measuring prescriptions is given in W. Heitler, *The Quantum Theory of Radiation* (Clarendon Press, Oxford, 1954), 3rd edition, p. 319.

[A-3] (p. 51). A spin-$\frac{1}{2}$ particle possessing an anomalous magnetic moment $\mu e/(2m)$ in addition to the normal Bohr magneton $e/(2m)$ gives rise to an interaction

$$\tfrac{1}{2}\mu\sigma_{\mu\nu}F_{\mu\nu}$$

in addition to the normal (or "minimal") interaction of Eq. [19.2]. Such a term which depends on the fields $F_{\mu\nu}$ rather than on the potentials Φ_μ is called a Pauli term. See W. Pauli, *Rev. Mod. Phys.* **13**, 203 (1941).

[A-4] (pp. 53, 59, 79). Sections 13, 14, and 18 are largely repetitions of Sections 5, 9, and 3, respectively. This is because Sections 1 to 12 were delivered by Pauli during the shorter summer term of 1950 while Section 13 started the winter term of 1950–1951 after the long summer vacations. In the German original this division is explicitely marked as Teil I and Teil II.

[A-5] (pp. 66, 73). Here there is a switch in the meaning of $\Psi(N_\lambda)$ from eigenvectors $|(N_\lambda)\rangle$ to amplitudes of these vectors in a general state Ψ. More correctly one should write

$$\Psi = \sum_{(N_\lambda)} |(N_\lambda)\rangle c(N_\lambda) .$$

Then it follows from the auxiliary conditions [15.14] and the equations at the top of p. 66 that

$$\sqrt{N_3+1}\,c(N_3+1, N_4) + \sqrt{N_4}\,c(N_3, N_4-1) = 0 ,$$
$$\sqrt{N_3}\,c(N_3-1, N_4) + \sqrt{N_4+1}\,c(N_3, N_4+1) = 0 .$$

Hence

$$c(N_3, N_4) = \delta_{N_3,N_4}(-1)^{N_3}c(0, 0) ,$$

so that with $c(0, 0) \neq 0$

$$|\Psi|^2 = \sum_{N_3,N_4} |c(N_3, N_4)|^2 = \infty .$$

The assertion on p. 73 is obtained as follows: The time-dependence of the norm of Ψ is given by

$$|\Psi(t)|^2 \equiv (\Psi^*(t), \eta\Psi(t)) = (\Psi^* \exp[-iH^\dagger t]\eta \exp[+iHt]\Psi)$$
$$= \sum_{(N_\lambda)} \sum_{(N'_\lambda)} c^*(N_\lambda)c(N'_\lambda)\langle(N_\lambda)|\exp[-iH^\dagger t]\eta \exp[+iHt]|(N'_\lambda)\rangle .$$

Now, from the equation on top of p. 72 it follows that

$$\exp[-iH^\dagger t]\eta = \eta \exp[-iHt] .$$

But from

$$\eta A^\dagger_\lambda = \varepsilon_\lambda A^\dagger_\lambda \eta ; \qquad \varepsilon_\lambda = \begin{cases} +1; & \lambda = 1, 2, 3 , \\ -1; & \lambda = 4 , \end{cases}$$

one has $[\eta, N_\lambda] = 0$ and

$$\langle(N_\lambda)|\eta|(N'_\lambda)\rangle = \delta_{(N_\lambda),(N'_\lambda)}(-1)^{N_4}\langle(0)|\eta|(0)\rangle .$$

Hence

$$|\Psi(t)|^2 = \sum_{(N_\lambda)} |c(N_\lambda)|^2(-1)^{N_4}\langle(0)|\eta|(0)\rangle$$

which is indeed constant in time and of the form indicated on p. 73.

[A-6] (p. 108). Here Pauli refers to unpublished work by R. Glauber "on the unitary operator which describes the time development of a system and the demonstrations of its connection with the 'in' and 'out' operators and that the S-matrix defined by means of those operators is indeed the same as Dyson's" (quoted from a letter by Roy J. Glauber to the Editor). This work is included in the paper by Yang and Feldman (Ref. 1 of Chapter 5) and acknowledged there in Footnote 14. Pauli had suggested to Yang that Glauber be made co-author of the Yang-Feldman paper.

[A-7] (p. 125). Dyson of course kept his promise, see *Phys. Rev.* **82**, 428 (1951); **83**, 608, 1207 (1951); *Proc. Roy. Soc.* (*London*) **A 207**, 395 (1951).

In Footnote 7 of the first of these papers which was submitted on December 11, 1950, Dyson thanks "Professor W. Pauli and Dr. Res Jost, who pointed out to him the necessity for working in terms of such field-averages" and refers to an unpublished calculation of Jost which "included many of the ideas of the present series of papers." This shows the importance the activity of the Zurich school had at the time these lectures were given by Pauli.

As to the case of spin-zero mentioned under 1 it was not possible to find out whether the analogous work has ever been done.

[A-8] (p. 139). The following comment to this Note has kindly been supplied by Klaus Hepp:

1. It can be proved that the idea of renormalization is correct in every order of perturbation theory. This has been pioneered by Dyson [1] and extended to various formalisms and on many levels of mathematical rigor by many authors.[2]

2. In certain models of interacting quantum fields in s-dimensional space-time (as in the Yukawa interaction

for $s = 2$,[3] the quartic boson self-interaction Φ^4 for $s = 3$,[4] and the Φ^3 interaction for $s = 4$[5]) the definition of a renormalized locally correct Hamiltonian can be given in a mathematically rigorous way. The (in the limit of no cut-off) infinite counter-terms are exactly those which are suggested by perturbation theory. For the Yukawa model the results of Glimm and Jaffe are very encouraging: the time evolution in the Heisenberg picture of the renormalized theory is local and leads to a consistent definition of the nonlinear terms in the quantum field equations.

[1] F. J. DYSON, *Phys. Rev.* **75**, 486, 1736 (1949).

[2] For references see the contributions of K. Hepp and H. Epstein in the *1970 Les Houches Lectures*, C. de Witt and R. Stora, editors (Gordon and Breach, New York, 1971), and O. STEINMANN, *Perturbation Expansions in Axiomatic Field Theory*, Lecture Notes in Physics (Springer, Berlin, 1971).

[3] J. GLIMM and A. JAFFE, *Ann. Phys.* (*N. Y.*) **60**, 321 (1970) and *J. Functional Analysis* **7**, 323 (1971).

[4] J. GLIMM, *Comm. Math. Phys.* **10**, 1 (1968).

[5] K. OSTERWALDER, ETH Thesis (Zurich, 1970).

Index

A CATALOG OF SELECTED

DOVER BOOKS

IN SCIENCE AND MATHEMATICS

THE HISTORICAL BACKGROUND OF CHEMISTRY, Henry M. Leicester. Evolution of ideas, not individual biography. Concentrates on formulation of a coherent set of chemical laws. 260pp. 5⅜ x 8½. 61053-5

A SHORT HISTORY OF CHEMISTRY, J. R. Partington. Classic exposition explores origins of chemistry, alchemy, early medical chemistry, nature of atmosphere, theory of valency, laws and structure of atomic theory, much more. 428pp. 5⅜ x 8½. (Available in U.S. only.) 65977-1

GENERAL CHEMISTRY, Linus Pauling. Revised 3rd edition of classic first-year text by Nobel laureate. Atomic and molecular structure, quantum mechanics, statistical mechanics, thermodynamics correlated with descriptive chemistry. Problems. 992pp. 5⅜ x 8½. 65622-5

Engineering

DE RE METALLICA, Georgius Agricola. The famous Hoover translation of greatest treatise on technological chemistry, engineering, geology, mining of early modern times (1556). All 289 original woodcuts. 638pp. 6¾ x 11. 60006-8

FUNDAMENTALS OF ASTRODYNAMICS, Roger Bate et al. Modern approach developed by U.S. Air Force Academy. Designed as a first course. Problems, exercises. Numerous illustrations. 455pp. 5⅜ x 8½. 60061-0

DYNAMICS OF FLUIDS IN POROUS MEDIA, Jacob Bear. For advanced students of ground water hydrology, soil mechanics and physics, drainage and irrigation engineering and more. 335 illustrations. Exercises, with answers. 784pp. 6⅛ x 9¼. 65675-6

ANALYTICAL MECHANICS OF GEARS, Earle Buckingham. Indispensable reference for modern gear manufacture covers conjugate gear-tooth action, gear-tooth profiles of various gears, many other topics. 263 figures. 102 tables. 546pp. 5⅜ x 8½. 65712-4

MECHANICS, J. P. Den Hartog. A classic introductory text or refresher. Hundreds of applications and design problems illuminate fundamentals of trusses, loaded beams and cables, etc. 334 answered problems. 462pp. 5⅜ x 8½. 60754-2

MECHANICAL VIBRATIONS, J. P. Den Hartog. Classic textbook offers lucid explanations and illustrative models, applying theories of vibrations to a variety of practical industrial engineering problems. Numerous figures. 233 problems, solutions. Appendix. Index. Preface. 436pp. 5⅜ x 8½. 64785-4

STRENGTH OF MATERIALS, J. P. Den Hartog. Full, clear treatment of basic material (tension, torsion, bending, etc.) plus advanced material on engineering methods, applications. 350 answered problems. 323pp. 5⅜ x 8½. 60755-0

A HISTORY OF MECHANICS, René Dugas. Monumental study of mechanical principles from antiquity to quantum mechanics. Contributions of ancient Greeks, Galileo, Leonardo, Kepler, Lagrange, many others. 671pp. 5⅜ x 8½. 65632-2

Physics